Oxford Physics Series

General Editors

E.J. BURGE D.J.E. INGRAM J.A.D. MATTHEW

Oxford Physics Series

---

1. F.N.H. ROBINSON: *Electromagnetism*
2. G. LANCASTER: *D.c. and a.c. circuits*
3. D.J.E. INGRAM: *Radiation and quantum physics*
5. B.R. JENNINGS and V.J. MORRIS: *Atoms in contact*
6. G.K.T. CONN: *Atoms and their structure*
7. R.L.F. BOYD: *Space physics;the study of plasmas in space*
8. J.L. MARTIN: *Basic quantum mechanics*
9. H.M. ROSENBERG: *The solid state*
10. J.G. TAYLOR: *Special relativity*
11. M. PRUTTON: *Surface physics*
12. G.A. JONES: *The properties of nuclei*
13. E.J. BURGE: *Atomic nuclei and their particles*
14. W.T. WELFORD: *Optics*
15. M. ROWAN-ROBINSON: *Cosmology*
16. D.A. FRASER: *The physics of semiconductor devices*
17. L. MACKINNON: *Mechanics and motion*

L. MACKINNON

Reader in Physics, University of Essex

Mechanics and motion

Clarendon Press, Oxford

Oxford University Press, Walton Street, Oxford OX2 6DP
OXFORD LONDON GLASGOW NEW YORK
TORONTO MELBOURNE WELLINGTON CAPE TOWN
IBADAN NAIROBI DAR ES SALAAM LUSAKA ADDIS ABABA
KUALA LUMPUR SINGAPORE JAKARTA HONG KONG TOKYO
DELHI BOMBAY CALCUTTA MADRAS KARACHI

Casebound ISBN 0 19 851825 0

Paperback ISBN 0 19 851843 9

Printed in Great Britain by
Thomson Litho Ltd, East Kilbride, Scotland.

Editor's foreword

In the revision of physics syllabuses at both school and university in the last decade or so classical mechanics gets less emphasis than in past eras, but teachers are rapidly rediscovering just how important the subject is in giving an essential background for appreciating currently more fashionable branches of physics. Dr. Mackinnon sets out to provide a short, simple mechanics text containing all the basic ideas and concepts which will be used both implicitly and explicitly in later parts of degree courses in physics and related disciplines. Though less comprehensive than many mechanics texts, *Mechanics and motion* succeeds in emphasizing the most important aspects of dynamics while avoiding undue mathematical complexity.

Mechanics and motion is a core text of the Oxford Physics Series covering material usually treated in the first year of honours courses in English universities or in the second year in Scottish honours courses. *Radiation and quantum physics*, *D.c. and a.c. circuits*, *Atoms in contact*, *Optics*, and *Electromagnetism* are further core texts, while second-level books in the Series such as *The solid state*, *Basic quantum mechanics*, and *Atomic nuclei and their particles* assume the kind of basic mechanics covered in Dr. Mackinnon's book.

November 1977

Preface

Physics is such a fascinating subject that the interested student can too easily be led into the temptation of probing into the subject more deeply than his conceptual understanding will allow. It is only too inviting to try to master atomic theory before mastering classical mechanics, with the consequence that one finds oneself out of one's depth; an understanding of the spin of an electron must follow, not precede, an understanding of angular momentum. This book is intended to provide the basic essential concepts. At times, it may perhaps exceed what is absolutely necessary; the account of the ellipsoid of inertia may not be essential, but the underlying concept of a second-order tensor is an important one. The short introduction to analytical classical mechanics not only compels thought about basic ideas, but also should allow the student to encounter the Hamiltonian before meeting it in quantum mechanics. It is important for the student to work through examples, and the short selection provided here should be regarded as no more than a bare minimum designed to help in achieving understanding.
I am grateful to Dr. J.A.D.Matthew for his helpful comments and to Mrs. M.F.Kimmitt for her efficient typing of the manuscript.

L. Mackinnon

Contents

1. The basis of classical mechanics

1.1. INTRODUCTION

The scientific study of movement is the proper concern of physics. Whether the study is of large objects, such as planets orbiting the sun, or of small objects, such as atoms vibrating in a solid or atomic particles colliding in a cloud or bubble chamber, the same basic principles, the laws of mechanics, apply. Such scientific laws are a useful interpretation of the results of experimental observations, useful because they can be applied with confidence both to predict and to interpret.

The need for experiment to derive such laws was not always understood. Over two thousand years ago, Aristotle, who was no experimentalist, taught as laws of mechanics both that a force was always necessary to keep a body in motion and also that a heavier object would always fall faster to the ground than a light object. The more practical and experimentally-minded Archimedes, on the other hand, correctly discovered the laws of levers and of machines, and founded the science of hydrostatics. By the sixteenth century, and not really until then, the need for experiment and quantitative study had become generally understood; Leonardo da Vinci had rekindled interest in the study of mechanics, and Stevinus and Galileo had shown that Aristotle was wrong. The stage was set for the development of classical mechanics as we now know it, and this development came very quickly.

1.2. NEWTON'S LAWS OF MOTION

In 1687, Isaac Newton published his book *Philosophiae Naturalis Principia Mathematica*, usually referred to as the *Principia*; this book has been called the greatest scientific work ever produced by the human intellect. Among the important ideas contained in it is the formulation of laws of motion, which were beginning to be understood in Newton's time from experimental studies and which Newton then clarified and codified. These three laws may be stated as follows:

Law 1. Every body continues in its state of rest or of uniform motion in a straight line, unless it is acted on by an external force.

Law 2. When a force acts on a body, the rate of change of momentum of the body with time is proportional to the magnitude of the force and takes place in the direction in which the force is acting.

Law 3. Action and reaction are equal and opposite.

Every student of physics should know and understand these fundamental laws, not just knowing the form of words but understanding their ultimate significance and the concepts which they contain.

The first important concept to consider is that of force. One does not need to be a physicist to feel that one understands the term 'mechanical force'. However, when one tries to relate this understanding to the first of Newton's laws, it may be found that one has to think a little more deeply. One example will suffice to illustrate this point. Someone is trying to push a heavy piece of furniture and it does not move. The would-be pusher is aware that he is applying a force, yet the state of rest of the furniture continues despite Newton's first law. Reconciliation with this law is not achieved until the frictional force between furniture and floor is taken into account; this frictional force is as real a force on the furniture as is the applied force. It is then that appreciation of forces as directed or vector quantities follows, and it is understood that change in motion of the furniture will occur only when the vector sum of all the forces acting is other than zero. A separate but equally important point to appreciate is that whenever a body does change its motion, then a force is present.

Very early on in the study of physics, one learns to distinguish between scalar and vector quantities. Recent practice in the teaching of mathematics has also been to encourage thinking about vectors and their manipulation at an early stage. At a more advanced level, in order to handle mathematical expressions involving vector quantities conveniently, the physicist of today has adopted the use and the notation of vector algebra, which is a useful form of mathematical shorthand. It is neccesary to distinguish clearly between vector and scalar quantities; in this book (contrary to the usual convention), an underline will be used to distinguish the vector quantity $\underline{a}$ from the scalar a. Therefore, force as a vector quantity will often be represented by the symbol $\underline{F}$

The next important concepts are those of momentum and of mass. The vector quantity momentum ($\underline{p}$) can be defined from

$$\underline{p} = m\underline{v},$$

where m is the mass, a scalar, and $\underline{v}$ is the velocity with which the mass is moving. If a force $\underline{F}$ acts on a mass m, then it follows from Newton's second law that

$$\underline{F} = \frac{\mathrm{d}}{\mathrm{d}t}(m\underline{v}); \tag{1.1}$$

provided now that m does not depend on $\underline{v}$, it will follow that

$$\underline{F} = m\frac{\mathrm{d}v}{\mathrm{d}t}$$

or (1.2)

$$\underline{F} = m\underline{a}$$

where $\underline{a}$ is the acceleration undergone by m as a result of the force $\underline{F}$. Eqn (1.2) is the equation of motion representing the useful practical consequence of Newton's second law; eqn (1.1) represents the law itself.

Clearly, if one can understand the concept mass, the understanding of momentum follows. The logical development of the science of mechanics has included considerable study of the meaning of mass; at this stage, it is convenient to accept Newton's idea that the mass of a body measures the quantity of matter in that body. Using this concept, Newton also put forward in the *Principia* his law of gravitation, which states that, when two masses are separated, they will attract one another with a force whose magnitude is proportional to the product of their masses and inversely proportional to the square of the distance between them. This can be expressed by the equation

$$F = G\frac{m_1 m_2}{r^2}, \tag{1.3}$$

where F is the magnitude of the force between two masses m_1 and m_2 separated by a distance r. This equation can now be used to get a measure of mass. Newton proved that the spherical earth acted gravitationally as though its entire mass M were concentrated at its centre. From eqn (1.3), by substituting M for m_1, the mass to be measured m for m_2, and the radius of the earth R for r, it follows that the weight W (in units of force) of the mass m is given by

$$W = G\frac{Mm}{R^2}. \tag{1.4}$$

It therefore follows, from eqn (1.4), that weights and masses are proportional. It was not absolutely necessary to use Newton's law of gravitation to make this point. It would have been adequate to have assumed the experimental result that the magnitude of the acceleration due to gravity, g, was the same for all falling objects at the earth's surface (in a vacuum, of course), and thus restated eqn (1.2) in the form

$$W = mg,$$

showing immediately that W and m are proportional to each other.

In the scientific literature, the mass to be used in eqn (1.2) is sometimes termed the inertial mass, and that to be used in eqn (1.3), the gravitational mass. No experiment has yet found any difference between these two masses.

It is normal to assume, as Newton did, that the mass of a body is not a function of its velocity. For the great majority of physical observations, this appears true within the accuracy of experimental measurement, and the assumption is therefore justified. However, experiment shows (as theory in fact predicts) that the mass of a body does depend on its velocity. It is found that if a body is moving with speed v, its mass m is given by

$$m = m_0\left(1 - \frac{v^2}{c^2}\right)^{-\frac{1}{2}}, \tag{1.5}$$

where c is the velocity of light and m_0 is the mass when the body is stationary, known also as the rest mass. Eqn (1.5) shows that one needs very high velocities indeed before m differs appreciably from m_0. However, such high velocities can be and are regularly achieved by the particles produced by the accelerating machines of nuclear physics and by the particles of the cosmic radiation (particles moving freely in outer space). Eqn (1.5) can thus have practical application and may have to be used even if its theoretical origin is not understood by the user.

1.2.1. *Mach's definition of mass*

Towards the end of the nineteenth century. Ernst Mach, one of the most thorough critical investigators of the whole field of classical mechanics, defined mass in a manner independent of the concept of quantity of matter. His definition can be stated as follows: when two free bodies of equal mass act on one another at a distance, they will produce in each other equal and opposite accelerations; if their masses are not equal, then the magnitude of these accelerations is in inverse ratio to their masses.

In this way, relative masses are defined using only the measurable quantity 'acceleration', and the term 'quantity of matter' is no longer needed. The validity of Newton's laws remains unaltered.

1.3. D'ALEMBERT'S PRINCIPLE OF INERTIAL REACTION

Eqn (1.2) stated

$$\underline{F} = m\underline{a}.$$

Thus $m\underline{a}$ is dimensionally a vector quantity analogous to force. The equation may be rewritten in the form

$$\underline{F} - m\underline{a} = 0. \tag{1.6}$$

Consider an object being accelerated by a force. Consider now everything from the point of view of the object, thus imagining the object to be stationary from its own point of view. If it is stationary, the total force on it must be zero. If $-m\underline{a}$ is considered to be a force, then eqn (1.6) gives the total force, states it to be zero, and, from the object's point of view, an equation in dynamics has become an equation in statics. The apparent force $-m\underline{a}$ is called the inertial reaction, and the substitution of it is sometimes called D'Alembert's principle. (In some ways, this is no more than a restatement of Newton's third law of motion.)

It might be felt that treating $-m\underline{a}$ as a force is somewhat artificial and no more than a mathematical device; a simple example will show that it is more than this. When an aircraft is accelerating during take-off, a passenger feels himself to be pushed against the back of his seat, which is itself pushing him from behind. From his point of view, he consciously feels two

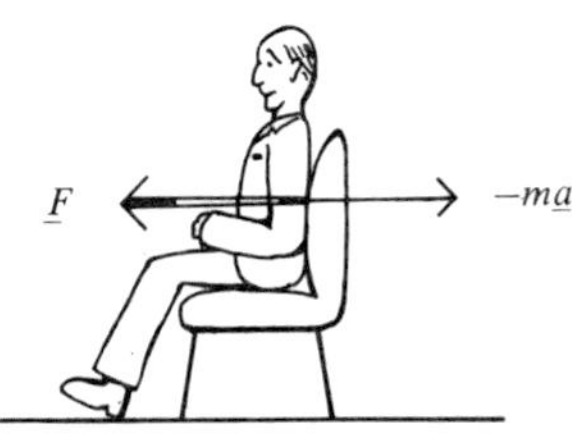

FIG. 1.1. To illustrate D'Alembert's principle; as the aircraft accelerates forward, the passenger feels himself pushed against the seat by the force, $-m\underline{a}$

forces. One is the force accelerating him, $\underline{F}$, and the other is the inertial reaction, $-m\underline{a}$, which D'Alembert's principle indicates to be the equivalent of a force (see Fig. 1.1). To him, however, this second force is every bit as real as the first. Another example is provided by the sinking feeling felt by lift passengers when the lift accelerates rapidly upwards. There is therefore nothing artificial about the force of inertial reaction; Newton's third law leads one to expect it.

1.4. KINEMATIC EQUATIONS FOR UNIFORM ACCELERATION

If a body is moving with uniform acceleration a, and its velocity increases from u to v over a time interval t, then a velocity - time graph will take the form shown as Fig. 1.2. The distance s travelled by the body during the period t is given by the shaded area (the integral of velocity times time element.)

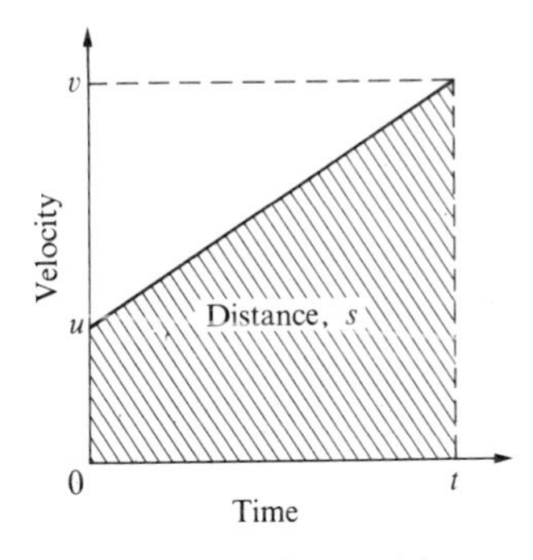

FIG. 1.2. Velocity - time graph for uniform acceleration.

As a is uniform, the graph is a straight line and one can write down at once that

$$a = \frac{v - u}{t}, \tag{1.7}$$

and

$$s = \left(\frac{v + u}{2}\right)t. \tag{1.8}$$

The product of the left-hand sides of eqns (1.7) and (1.8) equated to the product of the right-hand sides gives

$$2as = v^2 - u^2, \tag{1.9}$$

and substituting in eqn (1.8) for v using eqn (1.7) gives

$$s = ut + \tfrac{1}{2} at^2. \tag{1.10}$$

These four equations (1.7, 1.8, 1.9, and 1.10) are useful enough to be memorized.

1.5. SCALARS AND VECTORS

When dealing with problems in mechanics, it is useful to be able to distinguish between scalar and vector quantities. Scalar quantities, having magnitude and no direction, include mass and energy; vector quantities, having magnitude and direction, include force, velocity, acceleration, and momentum. Vector quantities also follow the rules of vector algebra, which is of considerable mathematical convenience as a form of shorthand notation and is therefore widely used by physicists. It may be noted in passing that the scalar magnitude of velocity is often called the speed. It is assumed here that the reader has some knowledge of this notation and of how to handle vector quantities; a forewarning of this point was made in Section 1.2.

1.6. UNITS AND DIMENSIONS

For all scientific work, an internationally agreed system of units (SI units) has been adopted. The kilogram is the basic unit of mass, and a force of one newton will give a mass of one kilogram an acceleration of one metre per second per second. As the acceleration due to gravity at the earth's surface is about $9{\cdot}8\ \mathrm{m\,s^{-2}}$, it follows that the force of weight of a 1 kg mass is about 9·8 N. So, if one desires to visualize orders of magnitude, lifting a weight of about 100 g (not far below the weight of a quarter-pound packet of sweets) will require a force of about 1 N. The dimensions of force are, from eqn (1.2), those of mass times acceleration, i.e. $\mathrm{MLT^{-2}}$. It should be noted that by defining force in terms of its effect on a mass, no dimensional constant appears in the equations of motion, eqns (1.1) and (1.2); such a constant has therefore to appear in eqn (1.3), which might otherwise itself have been used to define force in a very different way.

1.7. IMPULSE

Adopting the symbol $\underline{p}$ for momentum, eqn (1.1) can be rewritten as

$$\underline{F} = \frac{\mathrm{d}\underline{p}}{\mathrm{d}t} .$$

Therefore, if $\underline{p} = \underline{p}_0$ at $t = 0$ and $\underline{p} = \underline{p}_t$ at $t = t$, it follows that

$$\int_0^t \underline{F}\,\mathrm{d}t = \int_{\mathrm{p}_0}^{\mathrm{p}_t} \mathrm{d}\underline{p} = \underline{p}_t - \underline{p}_0. \qquad (1.11)$$

The integral on the left-hand side of eqn (1.11) is known as the impulse, and the equation itself tells one that impulse equals change of momentum. If $\underline{F}$ is not a function of time, the impulse becomes simply the product of $\underline{F}$ and the time for which it has acted. It is a vector quantity, and it is not usually represented by any particular symbol even though it is frequently useful when dealing with certain types of problem.

1.8. CONSERVATION OF MOMENTUM

An example of a problem to which the concept of impulse may be applied is a simple collision between two spherical objects travelling in the same straight line. As shown in Fig. 1.3(a), let the two objects have masses m_1 and m_2; before collision, let their velocities be $\underline{u}_1$ and $\underline{u}_2$ ($\underline{u}_1 > \underline{u}_2$ of course), and after

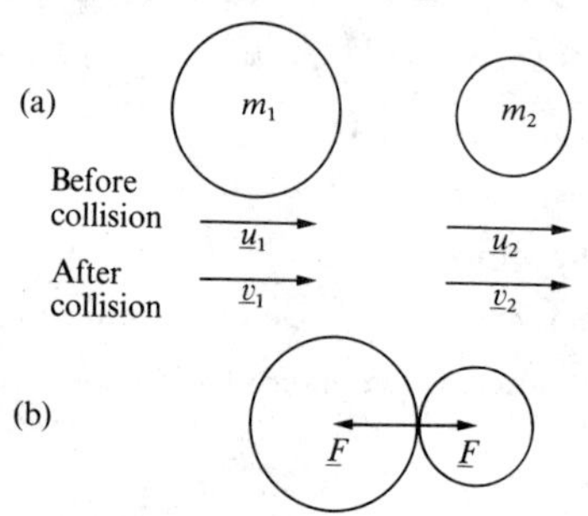

FIG. 1.3. To illustrate a simple collision between two spherical masses; (a) before and after collision, (b) during impact, with action and reaction always equal and opposite.

collision, $\underline{v}_1$ and $\underline{v}_2$ respectively. When they collide, they will be momentarily in contact, as shown in Fig. 1.3(b), and there will be a force between them, each experiencing a force from the other. By Newton's third law, these forces will be equal and opposite throughout the entire period of contact, t. The impulse equation, derived from Newton's second law as eqn (1.11), becomes for mass m_1

$$-\int_0^t \underline{F}\,\mathrm{d}t = m_1\underline{v}_1 - m_1\underline{u}_1, \tag{1.12}$$

and for mass m_2

$$\int_0^t \underline{F}\,\mathrm{d}t = m_2\underline{v}_2 - m_2\underline{u}_2. \tag{1.13}$$

From eqns (1.12) and (1.13), the fact that the impulses are equal and opposite leads at once to

$$-(m_1\underline{v}_1 - m_1\underline{u}_1) = m_2\underline{v}_2 - m_2\underline{u}_2 \tag{1.14}$$

or

$$m_1\underline{u}_1 + m_2\underline{u}_2 = m_1\underline{v}_1 + m_2\underline{v}_2. \tag{1.15}$$

Eqn (1.15) states that the total momentum before the collision equals the total momentum after the collision. In other words, no momentum has been either gained or lost as a result of the collision. The problem is a very simple one, and the whole discussion has been simplified by keeping all the motion in one direction. The same conclusion would however have been reached, i.e., that there is no change in total momentum if Newton's laws are obeyed, had the problem been more elaborate, provided that no outside source of forces had been allowed to act on the system under discussion. This leads to an important general principle, namely that of the conservation of momentum, which may be stated as follows;

'In any isolated system, the momentum remains constant.'

There is no known exception to this rule. An example of an isolated system is provided by the rocket discussed in Section 1.15.

1.9. WORK AND ENERGY

The concepts of work and of energy are fundamental to the explanation of many physical phenomena; to state that physics is all about energy would hardly satisfy as a definition of physics, but it is a statement with a lot of truth in it.

Work is simply defined as follows. When a force $\underline{F}$ is allowed to act through a distance $\underline{s}$, then the work done is $\underline{F}\cdot\underline{s}$ (i.e., the scalar product of $\underline{F}$ and $\underline{s}$). Work is thus a scalar quantity.

The SI unit of work is the joule (1 N acting through 1 m), and the obsolete cgs unit is the erg (1 dyne acting through 1 cm).

Energy is defined by the capacity to do work. The energy of a body is simply the amount of work that can in principle be got from the body. Kinetic energy is energy held as a result of motion and potential energy is energy held by virtue of position in a force-field. The term 'potential' used to describe a force-field very often simply means potential energy per unit quantity: in an electric field, the potential is the potential energy per unit charge, and in a gravitational field, the potential is the potential energy per unit mass. Potential does not always quite mean this, but potential energy is quite explicitly defined. In symbols, it is convenient to use W for work, E for energy in general, T for kinetic energy and V for potential energy.

If a body of mass m is moving with velocity v, then its kinetic energy T can be readily deduced, using Newton's laws of motion and a simple kinematic equation, as follows. Let the body be brought to a halt by a uniform force acting on it over a distance s. The body will act against this uniform force, by Newton's third law, with an equal and opposite force F, and so do work Fs as it stops. The kinetic energy T of the body, being the work that can be got from the body as a result of its motion, is thus given by

$$T = Fs. \tag{1.16}$$

Suppose that the time over which the uniform stopping force, $-F$ (minus because it is opposite to F) has acted is t. The momentum has changed by $-mv$. So impulse considerations give

$$(-F)t = -mv$$

or

$$F = \frac{mv}{t} \tag{1.17}$$

Applying the kinematic equation, eqn (1.8), to this problem, one obtains

$$s = \frac{vt}{2} \tag{1.18}$$

Substituting into eqn (1.16) for F from eqn (1.17) and for s from eqn (1.18) gives

$$T = \left(\frac{mv}{t}\right)\left(\frac{vt}{2}\right),$$

i.e.,

$$T = \frac{1}{2}mv^2. \tag{1.19}$$

This result is, of course, very well known and quite fundamental. T will always be positive; the sign or direction of v does not affect it. Energy, like work, is a scalar quantity.

1.10. POWER

Power is simply rate of doing work. If work W is done in time t, then the power P is W/t. The SI unit of power is the watt, W. Watts are equivalent to joules per second, J s^{-1}.

1.11. EQUIVALENCE OF MASS AND ENERGY

One result of the special theory of relativity has already been quoted as eqn (1.5), i.e., that the mass m of a body is related to its velocity v by

$$m = m_o\left(1 - \frac{v^2}{c^2}\right)^{-\frac{1}{2}}, \tag{1.20}$$

where m_o is the rest mass and c is the velocity of light. As c is about 186 000 miles per second (i.e., around 3×10^8 m s^{-1}), it is reasonable to suppose that in most cases $v^2 \ll c^2$; thus one can expand the right-hand side of eqn (1.20) by the binomial theorem and write

$$m = m_o(1 + \frac{1}{2}\frac{v^2}{c^2} + \text{smaller terms}).$$

Neglecting the smaller terms and multiplying both sides by c^2, it follows that

$$mc^2 = m_oc^2 + \frac{1}{2}m_ov^2. \qquad (1.21)$$

The kinetic energy of the body is immediately recognizable; furthermore eqn (1.21) makes it clear that mc^2 and m_oc^2 have the dimensions of energy. In fact, we may regard mass as if it were a form of energy; units of mass have only to be multiplied by the dimensional constant c^2 to become units of energy. Einstein's equation relating energy E to mass m, namely

$$E = mc^2,$$

can be seen to follow naturally.

This equivalence of energy and mass is often important when considering the mechanics of atomic processes. The scale is such that a mass of one gram is equivalent to 9×10^{10} joules; thus only a small mass change, in mass units, is equivalent to a large energy change, in energy units. The generation of nuclear power arises from just such a mass to energy change, or, to state it more correctly, from the exchange of energy in the form of mass into other forms of energy.

1.12. CONSERVATION OF ENERGY

Consider again the collision process illustrated in Fig. 1.3 and discussed in Section 1.8. Suppose that during the collision the interacting force F acts through a distance s. Suppose further that the system, i.e., the two spherical objects travelling in the same straight line, can only hold energy as kinetic energy; in general, this supposition would not be justified but in certain cases, such as certain atomic collisions, it would be reasonable. If, however, it is assumed to be true in this case, then all the work done by or on the interacting force F will be converted to kinetic energy. During the collision, mass m_2 will act on m_1, slowing it down, so that

$$\frac{1}{2}m_1u_1^2 - \frac{1}{2}m_1v_1^2 = \int_0^s F ds. \tag{1.22}$$

Similarly, m_1 will be acting on m_2, only this time

$$\frac{1}{2}m_2v_2^2 - \frac{1}{2}m_2u_2^2 = \int_0^s F ds. \tag{1.23}$$

It is worth emphasizing that eqns (1.22) and (1.23) can only hold true if all the work done by the interacting force is convertible to kinetic energy. If this is so, the left-hand sides can be equated to give

$$\frac{1}{2}m_1u_1^2 - \frac{1}{2}m_1v_1^2 = \frac{1}{2}m_2v_2^2 - \frac{1}{2}m_2u_2^2. \tag{1.24}$$

This equation can be rearranged to give

$$\frac{1}{2}m_1u_1^2 + \frac{1}{2}m_2u_2^2 = \frac{1}{2}m_1v_1^2 + \frac{1}{2}m_2v_2^2. \tag{1.25}$$

Eqn (1.25) is a straightforward statement that the kinetic energy before the collision equals the kinetic energy after the collision, i.e., that kinetic energy has been conserved during the collision.

Suppose now that such a collision has taken place. The conservation of momentum indicates from eqn (1.14) that

$$m_1(u_1 - v_1) = m_2(v_2 - u_2), \tag{1.26}$$

and the conservation of energy indicates from eqn (1.24) that

$$m_1(u_1^2 - v_1^2) = m_2(v_2^2 - u_2^2). \tag{1.27}$$

Eqn (1.27) can be rewritten as

$$m_1(u_1 - v_1)(u_1 + v_1) = m_2(v_2 - u_2)(v_2 + u_2),$$

so that, using eqn (1.26)

$$u_1 + v_1 = v_2 + u_2;$$

this last equation can be rearranged to give

$$\frac{v_2 - v_1}{u_1 - u_2} = 1. \qquad (1.28)$$

This equation can be tested experimentally (as can the others). Such studies had been carried out on large-sized objects (i.e., large by atomic standards!) by Newton's time, and Newton found that the ratio $(v_2 - v_1)/(u_1 - u_2)$ was in general less than 1. The ratio has been given a name, coefficient of restitution, e; and in general $0 \leqslant e \leqslant 1$. This quantity is not itself usually of any great importance to a physicist; it has really become of historical interest only. If $e = 1$, the argument leading to eqn (1.28) shows that kinetic energy has not been lost over the collision, which is then said to be perfectly elastic. Elastic collision processes are frequent realities on the atomic scale, but rare on the large scale. Large-scale collisions are usually not elastic; the purist may reserve the term inelastic for $e = 0$, but inelastic may in fact mean any collision for which $e < 1$, i.e., any collision during which some kinetic energy seems to have disappeared. (One can hardly overemphasize the fact that one does not use the principle of the conservation of energy when treating a collision process unless one is absolutely certain that the collision is perfectly elastic.)

The fact that real collisions were not elastic, so that energy seemed to be lost during a collision process, meant that for many years, energy as such did not seem to have the significance that an obviously conserved quantity would have had. The significance really began to be appreciated when the experimentalists of the nineteenth century showed that mechanical work could be converted into heat; following this came the understanding that heat itself was a manifestation of molecular motion. Motion carries energy, so heat itself is naturally a form of energy. The fact that a given amount of work could be converted to a calculable amount of heat led on to the principle of the conservation of energy, first stated in the mid-nineteenth century. This is a most important principle; simply stated, it is: 'energy can neither be created or destroyed, but can be converted from one form to another'.

In an inelastic collision, it can be shown that the energy apparently lost during impact has nearly all been converted to heat in the colliding bodies; if the collision was in air, a little of the energy may have been converted to sound. That this can happen when the colliding bodies are large is due to the fact that the bodies are made up of atoms; a collision sets these atoms into further vibrations in the solid, evidenced as heating of the solid. An isolated atom, on the other hand, may have no easy way of taking up energy as internal motion; collisions between such atoms have therefore to be elastic, conserving both kinetic energy and momentum. In the theory of collision processes on the atomic scale, full account has to be taken of both energy and momentum; if kinetic energy is not conserved, its disappearance has to be accounted for; where it has gone may be crucial to the theory, and to the experiment.

The mechanism whereby mechanical work is converted to heat is very often that of friction. The stopping of the moving body of Section 1.9 is likely to be a frictional process, producing heat at the rubbing surfaces; it is easy to visualize the mechanical processes involved here on an atomic scale. The conversion of kinetic energy to heat in an inelastic collision may be attributed to internal friction in the colliding objects. This internal friction prevents the deformations, which occur as a result of the collision, from being truly elastic; heat is generated by internal atomic movements during deformations and the bodies may fail to return immediately to their original shape after the impact is over. Internal friction processes can be very important technologically; metal fatigue is an example of such a process. In this case, the mechanical properties of the metal alter following changes in its structure brought about by the atomic movement accompanying repeated deformations.

To sum up the preceding discussion about collisions, the important points are these. The conservation of momentum can always be applied, whereas the conservation of energy can only be usefully applied to perfectly elastic collisions or to processes where it is possible fully to account for possible energy dissipation. To solve a collision problem completely does however require more knowledge than momentum conservation gives; the kinetic energy loss has to be known if the behaviour after the collision is to be predicted. A knowledge of the coefficient of restitution suffices, in that it implicitly takes into account the kinetic energy loss; however, the coefficient of restitution is itself not always a constant, so one has to be wary of it.

1.13. COLLISIONS IN OTHER THAN A STRAIGHT LINE

The collision problem so far discussed and illustrated by Fig. 1.3 is limited in that, being a 'head-on' collision, all motion is in one dimension only. A real collision process is not likely to be so limited. Fig. 1.4. indicates a more likely situation. In such circumstances, the problem can be tackled theoretically by remembering that momentum is a vector quantity, so that all momentum changes take place only in the components of momenta whose direction is that of the interacting force. The whole problem becomes then a simple one of resolution into

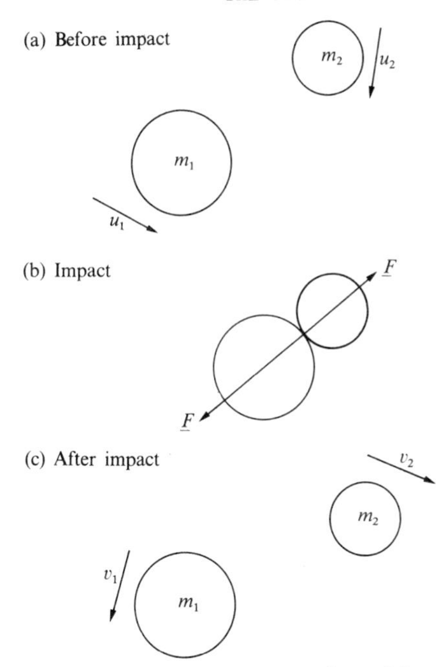

FIG. 1.4. To illustrate the more general problem of a simple collision than is shown by Fig. 1.3, the problem being no longer unidimensional.

components. Again, knowledge of the elastic nature of the collision, or of the energy loss, is necessary for a complete solution.

1.14. CENTRE OF MASS DURING COLLISION

Return now to the simple unidimensional collision shown in Fig. 1.3 and consider the effect of the collision on the centre of mass of the system. Fig. 1.5 illustrates this event again in a form suitable for this discussion. Let point P be the centre of mass of the system. For this to be so,

$$m_1 x = m_2 y, \tag{1.29}$$

where x and y are the distances of m_1 and m_2 respectively from P.

It is assumed here that there is no need to derive eqn (1.29). One simple way of thinking of it is to imagine m_1 and m_2

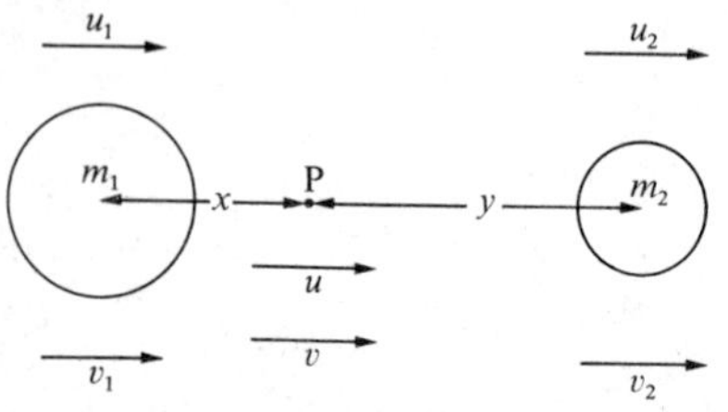

FIG. 1.5. To illustrate discussion of centre of mass movement.

connected by a light rod and then placed on a fulcrum; to achieve balance, this fulcrum would have to be located at the centre of mass. Simple moments then give eqn (1.29).

Eqn (1.29) must hold at all times. Therefore

$$m_1\frac{dx}{dt} = m_2\frac{dy}{dt}. \tag{1.30}$$

Let the centre of mass be moving with a velocity u before the collision and v after the collision. By inspection of Fig. 1.5, it is clear that, before the collision, eqn (1.30) takes form

$$m_1(u - u_1) = m_2(u_2 - u), \tag{1.31}$$

since (dx/dt) is negative and equal to $(u - u_1)$, etc. Similarly, after the collision,

$$m_1(v - v_1) = m_2(v_2 - v). \tag{1.32}$$

Eqns (1.31) and (1.32) can now be rewritten as

$$(m_1 + m_2)u = m_1u_1 + m_2u_2, \tag{1.33}$$

and

$$(m_1 + m_2)v = m_1v_1 + m_2v_2. \tag{1.34}$$

Conservation of momentum, i.e. eqn (1.15), shows that the right-hand sides of eqns (1.33) and (1.34) are equal; in which case, the left-hand sides too must be equal, so it follows that

$$u = v.$$

Thus the motion of the centre of mass is unaffected by the collision. The system consisting of the two masses m_1 and m_2 behaves throughout the collision as if it were a total mass $(m_1 + m_2)$ placed at the centre of mass and moving with constant velocity. A force external to the system would be necessary to change this state of affairs. This result will apply to any isolated system and is a natural consequence of the principle of the conservation of momentum.

1.15. MOTION RELATIVE TO CENTRE OF MASS

The occasion often arises when it is convenient to think of a process in terms of another point of view than one's own. When discussing D'Alembert's principle in Section 1.3, the feelings of a passenger in an accelerating aircraft were considered, both from the point of view of an observer on the runway who saw the problem as one of an accelerating passenger, and from the point of view of the passenger himself, who could see and feel it as a problem in statics. In the same sort of way, it is often convenient to consider a collision process from the point of view of the centre of mass of the system. Mathematically, this means transforming (or changing) the coordinates in which the process is described from a set fixed to the original observer of the process to a set fixed to the centre of mass of the system. Such a coordinate transformation can, in the case of collision processes, simplify the mathematical treatment of the problem and thus be well worthwhile.

Consider now the transformation of coordinates to the centre of mass coordinate system for the collision process discussed in Section 1.14 and illustrated by Fig. 1.5. The result of such a transformation is shown by Fig. 1.6, where m_1 and m_2 now have velocities before collision of u_1' and u_2' and after collision of

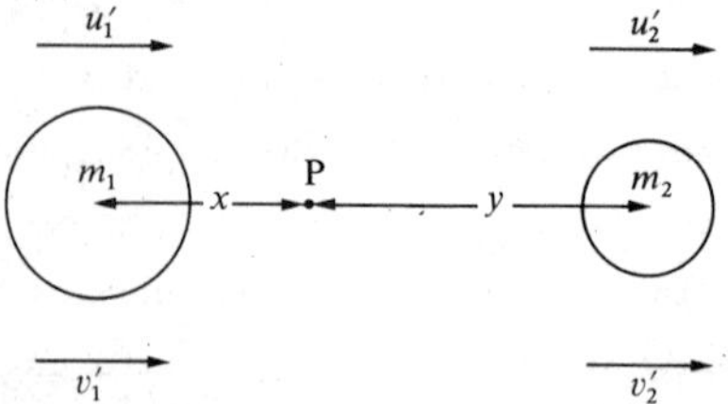

FIG. 1.6. This is the same as Fig. 1.5, but the velocities are now those relative to the centre of mass.

v_1' and v_2'. Eqn (1.30) tells one that

$$m_1\frac{dx}{dt} = m_2\frac{dy}{dt}.$$

Before the collision, $(dx/dt) = -u_1'$ and $(dy/dt) = u_2'$, so that

$$-m_1u_1' = m_2u_2'$$

or

$$m_1u_1' + m_2u_2' = 0. \tag{1.35}$$

Similarly, after the collision, it follows that

$$m_1v_1' + m_2v_2' = 0. \tag{1.36}$$

Eqns (1.35) and (1.36) make the important point that, in the centre of mass coordinate system, the total momentum is always zero. This result really follows almost intuitively from the discussion of Section 1.14.

An example of the type of problem for which consideration of the centre of mass motion is useful is provided by a rocket, well out in space, whose course is to be altered. As the rocket is well out in space, it will be effectively free from outside forces (i.e., gravitational forces) and it plus its load (of fuel, etc.) forms an isolated system, whose centre of mass will continue to move with constant velocity. If the rocket is then to change course, mass (i.e., fuel) must be expelled from it. The resulting motion of the expelled mass and of the rocket will be such as to keep their centre of mass in unaltered constant velocity. If one wishes to calculate how much mass to expel, in which direction and with what velocity, reference to the centre of mass coordinate system, wherein the momentum of the system remains zero, tells one that, in that system, the expelled mass must carry equal and opposite momentum to the momentum it is desired to impart to the rocket (in that system). So the calculation of how to achieve a desired change of course can be simplified.

The centre of mass coordinate system is also extremely useful when considering collision problems in atomic or nuclear physics. Such collisions are often elastic, so that kinetic energy is conserved; as velocities differ between coordinate systems, kinetic energies also differ. However, irrespective of which system of coordinates is used, conservation of kinetic energy will apply for elastic collisions. It turns out therefore that one can usefully employ a transformation to centre of mass coordinates when considering both energy and momentum.

1.16. INTERACTIONS OTHER THAN BY IMPACT

In the discussion so far, all the interactions have been simple collisions, where the interacting force lasted only a very short time. In many real cases of interest to physicists, the interactions last very much longer. The force between the interacting bodies may extend so as to be felt while the bodies are far apart. When this is so, the concept of potential can often be useful in dealing with the theory; this concept was briefly mentioned in Section 1.9. For convenience, it is customary to define the potential at a point in a gravitational or electrical field as the work done in bringing unit mass or charge to that point from infinity. Defining it in this way, potential or potential energy may well prove to be negative, but one cannot expect kinetic energy ever to be negative for any real observable object.

Chapter 5 deals with motion under a central force, which is one form of an interaction which is other than by impact. One of the first atomic collision problems likely to be encountered by a physics student is of this type. This is the scattering of fast-moving α-particles by atomic nuclei, a process known as Rutherford scattering. Here the interacting force is the repulsive electrostatic (coulomb) force between the positive charges on the two colliding particles, the α-particle and the nucleus, a force which obeys the inverse square law. The theory of the collision process is lengthier than in the case of simple impact (see Section 5.6); nevertheless, the basic mechanical principles, such as momentum conservation, remain valid and useful when applied with suitable care.

1.17. SUMMARY

In this chapter, the basic laws and concepts of mechanics, as formulated by Newton, have been reviewed, and certain important consequences thereof have been discussed. The principles of conservation of momentum and conservation of energy have been introduced, together with their application to very simple interaction processes. On the whole, this chapter has been concerned only with motion in a straight line. However, the usefulness of mechanics lies in its application to a wide range of physical problems. It is therefore important for a student to study how to carry out this application by working through examples. The limited number of examples, which now follow, have been chosen with a view to their instructive content, and no student should be satisfied until he or she can easily cope with them and other similar problems.

1.18. EXAMPLES

1. A horse is pulling a cart. Action and reaction are equal

and opposite, so that the cart always resists with the same force as the horse is pulling it. How then does the horse succeed in moving the cart?

2. Which of the following are forces of inertial reaction?
(a) The force with which a seat-belt holds a driver in place when a car stops suddenly.
(b) The force with which a bullet causes a flesh wound.
(c) The force that causes a china jug to break when dropped on to the floor.

3. A tall building has a height of 200 m between the ground floor and a restaurant on the highest floor. An express lift conveys passengers non-stop from the ground floor to the restaurant. If, for their comfort, passengers are not to experience an acceleration greater than $\pm 0{\cdot}2\ g$, what is the shortest possible time for the lift's journey? If the lift cage weighs 2000 kg and if the maximum load it is to carry is a further 3000 kg, calculate the maximum power expanded by the lift's motor in raising the lift. (The weight of cable may be assumed negligible for this calculation). When is this maximum power being used? (Assume $g = 9{\cdot}8\ \mathrm{m\,s^{-2}}$)

4. A car of mass M is travelling with velocity v. The brakes apply a force F but unfortunately fail after a time t. What is the maximum velocity the car could have had for it to have been brought to a halt before the brakes failed?

5. A fireman's hose has a jet which delivers $X\ \mathrm{m}^3$ of water of density ρ at velocity v in time t. The jet is directed horizontally on to a wall. What is the force on the wall? If the jet maintains a constant area of cross-section, what is the pressure it exerts over the area of contact with the wall?

6. A car of mass M is travelling with velocity v. The brakes apply a force F which obeys the rule

$$F = A + Bv$$

where A and B are constants. Show that the time t needed to bring the car to a halt is given by

$$t = \frac{M}{B} \ln \left(1 + \frac{B}{A} v\right) .$$

7. Show that if a car is brought to a halt by a constant braking force, the stopping distances at 75, 60, and 30 m.p.h. are in the ratio 6·2 : 4 : 1.

8. In aluminium, the fastest conduction electrons move freely with speeds about two million miles per hour. By what percentage do their masses exceed their rest masses? (Assume 1 km = 0·6 mile and $c = 3 \times 10^8$ m s^{-1}.)

9. A ballistic pendulum consists of a heavy block of wood of mass M on the end of a long light cord of length l which is held from a rigid support. A bullet of mass m ($<<M$) is fired horizontally into the block of wood. On impact, the bullet is imbedded and the mass M swings back a distance d. (a) Find an expression for the velocity v of the bullet in terms of d, etc. (b) Estimate the fraction of the kinetic energy lost on impact.

10. A mass m travelling with velocity v is slowed down by a head-on elastic collision with a second mass. What should be the magnitude of this second mass if the speed of m after collision is to be as small as possible?

11. Two spherical bodies each of mass m undergo a perfectly elastic collision, one being initially stationary and the other hitting it. Show that, for nearly all angles of impact, the paths with which the bodies move after the collision will be at right angles to one another.

2. Rotary motion

2.1. INTRODUCTION

The motion of a wheel, or of a spinning top, represents a rotary motion. Everything involved in the motion is rotating about an axis: a line in space. Rotations can also take place on a small scale; molecules, for example, can rotate. Thus the physicist (or chemist or engineer) has to be able to describe rotary motions with the same facility as he describes linear motions. This chapter is therefore devoted to the topic of rotary motion, and will start with a number of elementary definitions. Where it is applicable and useful, vector algebra will be used.

2.2. DEFINITIONS

Rotary displacement is measured in units of angle. Fig. 2.1 illustrates a rotary displacement; a uniform circular disc of

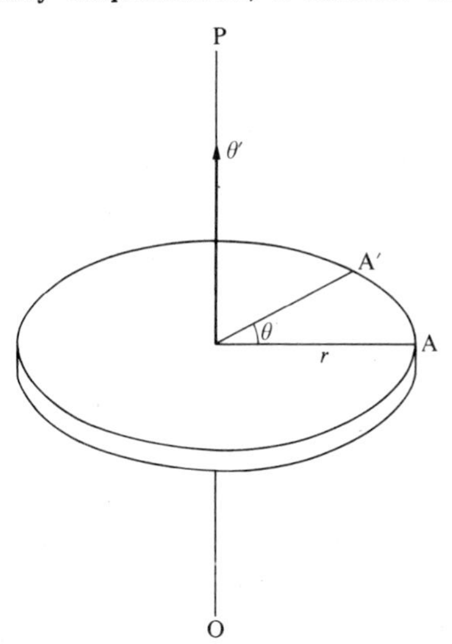

FIG. 2.1. To illustrate angular definitions; a uniform circular disc of radius r is rotated about the axis OP through its centre. It is moved through an angle θ.

radius r is moved round an axis OP through its centre so that the part of the disc originally at point A has moved to point A′. The magnitude of the angular displacement θ is then given by the ratio, (distance AA' measured round the circumference)/r, and is measured in radians.

In addition to magnitude, there is a direction associated with rotary displacement, namely the axis of rotation. It would be possible to represent the rotation through θ by a line, of length proportional to θ, along the axis; this line could be directed towards O or towards P. One can insist that it be towards P by specifying a right-handed corkscrew rule so that, if the observer looks from O and the motion A to A′ then appears clockwise, the line shall point away from him. If one does this, one gets the line θ' representing the displacement. In this way, one has given the displacement two of the qualities of a vector quantity, namely magnitude and direction. However, unfortunately θ' turns out not to be a vector. Two finite angular displacements do not add together according to the laws of vector algebra. A formal mathematical proof of this is too lengthy for this book, but it can be quickly proved experimentally as follows. Take a copy of this or any other book, and give it two right angle displacements in directions at right angles to one another; it will then be found that the final position of the book will depend on the order in which the two displacements were made. Had the displacements been vector quantities, this dependence on order would not have arisen.

This discussion of the directed nature of an angular displacement would be pretty pointless were it to end here. It does not. If, instead of being finite, the displacements are infinitesimal, then it can be shown that they do obey the laws of vector algebra and can be treated as vector quantities. Thus if one has an infinitesimal angular displacement $d\theta$ and if one then represents it by a vector $d\underline{\theta}$, defined as above, and follow this (or precede this) by a further similarly-defined displacement $d\underline{\phi}$ about some other axis, then the total displacement $d\underline{\psi}$ will be given both in magnitude and direction by

$$d\underline{\psi} = d\underline{\theta} + d\underline{\phi}. \tag{2.1}$$

This equation can also be proved formally, but the proof is beyond the scope of this book. The truly enquiring physicist may reasonably wish to seek out such a proof in a more advanced text just to satisfy himself; however, just as one uses a watch without knowing the details of its construction, so one can speed things up by accepting for the moment that the equation works.

The rotation speed $\omega (= d\theta/dt)$ is normally measured in radians per second; sometimes revolutions per unit time are quoted. The angular velocity $\underline{\omega}$ has the magnitude of the speed and is directed along the axis of rotation according to the corkscrew rule given above. As $\underline{\omega} = d\underline{\theta}/dt$, and $d\underline{\theta}$ is a vector quantity, it follows that $\underline{\omega}$ is a vector quantity. This also follows just by dividing eqn (2.1) by dt, when one obtains

$$\frac{d\underline{\psi}}{dt} = \frac{d\underline{\theta}}{dt} + \frac{d\underline{\phi}}{dt},$$

or

$$\underline{\omega}_3 = \underline{\omega}_1 + \underline{\omega}_2, \tag{2.2}$$

where the $\underline{\omega}$s are substituted as indicated and add vectorially;

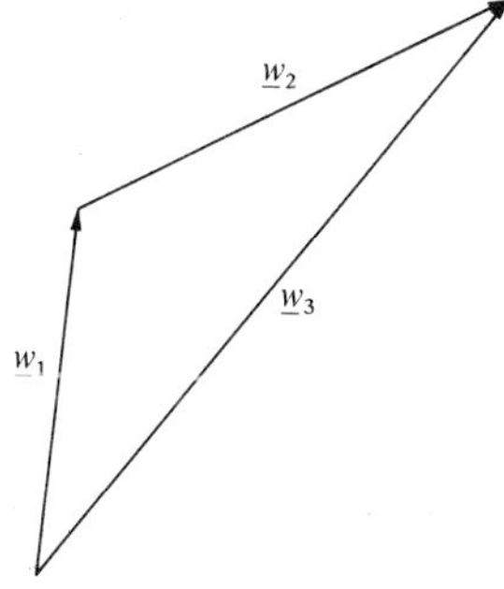

FIG. 2.2. The compounding of angular velocities as vectors; see eqn. (2.2).

Fig. 2.2 is intended to bring this point home.

Angular acceleration $\underline{\alpha}$ (or $\underline{\dot{\omega}}$) is clearly similarly a vector quantity. Angular accelerations, like linear accelerations, are brought about by the action of forces. The effective action of a force in producing motion around any point is called the moment of the force, or the torque, about that point.

For many students of physics, one of the earliest experiments to be carried out in a laboratory is an experiment to verify the law of the lever; from this experiment, the idea develops that the moment of a force is 'distance from fulcrum' times 'magnitude of force'. One then goes on to define moments as 'perpendicular distance to the line of action of the force' times 'magnitude of force'. One tends to treat the quantities so defined as if they were scalars, which is perfectly reasonable if one is only dealing with magnitudes. However, there are often occasions when directions have also to be taken into account; vector algebra provides a convenient method of so doing.

Using vector terminology then, the definition of a moment is as follows: the moment $\underline{T}$ of a force about a point is the vector product of the distance $\underline{r}$ from the point to the point where the force is acting and the force $\underline{F}$,

i.e.,
$$\underline{T} = \underline{r} \times \underline{F} \tag{2.3}$$

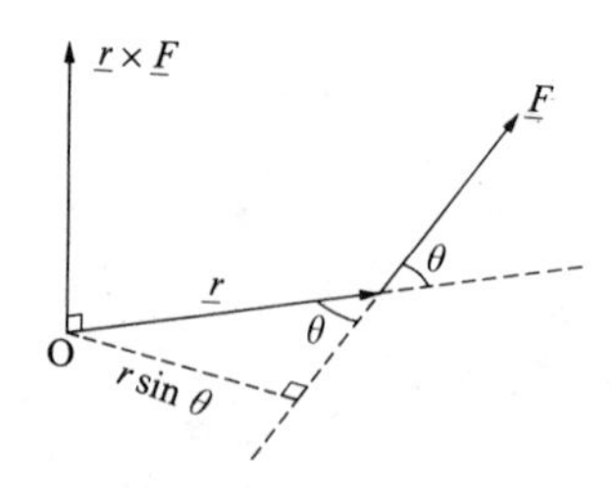

FIG. 2.3. To illustrate the definition of the moment of the force $\underline{F}$ about the point O. Note that the magnitude of the moment, $\underline{r} \times \underline{F}$, equals $rF\sin\theta$, or 'perpendicular distance from O to line of action of force' ($r\sin\theta$) times 'force' (F). Note also eqn (2.3).

This is illustrated by Fig. 2.3, where the relationship of this definition to the earlier definition is emphasized in the caption.

When a body is rotating, it is normally rotating about an axis; one may therefore wish to define a moment about an axis rather than about a point. As moment, or torque, is a vector quantity, to find the torque about an axis one only has to find the component, in the direction of the axis, of the total torque about any point on that axis. So, if the distance $\underline{r}$ and the force $\underline{F}$ have components $\underline{r}_{\parallel}$ and $\underline{F}_{\parallel}$ in the direction of the axis and $\underline{r}_{\perp}$ and $\underline{F}_{\perp}$ at right angles to the axis, it can readily be shown that the component of $\underline{T}$ parallel to the axis, $\underline{T}$, is given by

$$\underline{T}_{\parallel} = \underline{r}_{\perp} \times \underline{F}_{\perp} \qquad (2.4)$$

having magnitude simply $r\ F$. Fig. 2.4 illustrates this. In many cases, the axis of rotation will be that of the torque anyway, so

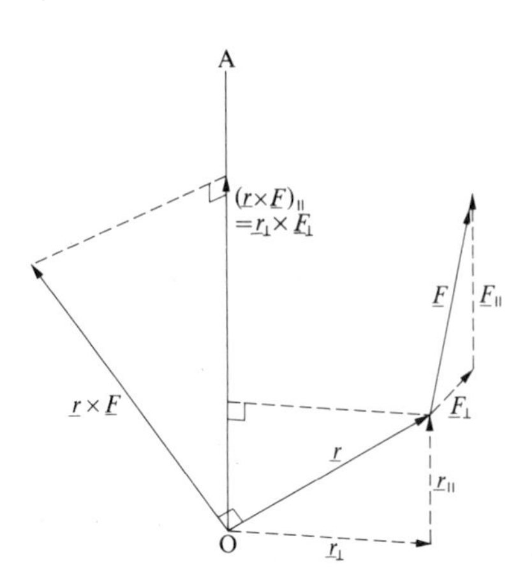

FIG. 2.4. To illustrate the relationship between torque about a point and torque about an axis (OA); the proof that $(\underline{r} \times \underline{F})_{\parallel} = \underline{r}_{\perp} \times \underline{F}_{\perp}$ is a simple matter of writing $\underline{r}$ and $\underline{F}$ in components, taking the products and inspecting the answer.

that eqn (2.4) will be identical in all respects to eqn (2.3).

2.3. CIRCULAR MOTION

Many of the principles inherent in the treatment of rotary motion can be illustrated by the motion of a particle in a circle. By considering all effects to be in the plane of the circle only, this becomes a two-dimensional problem.

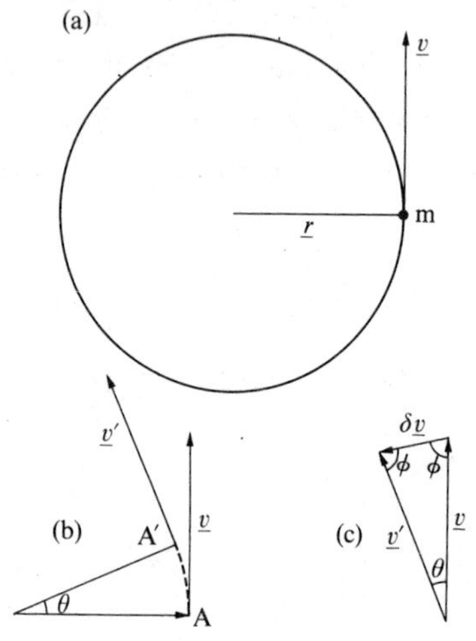

FIG. 2.5. To illustrate uniform motion in a circle.

Consider first the case of uniform motion as illustrated by Fig. 2.5. A particle of mass m is moving with uniform speed v around a circle of radius r. At any instant in time, the position of the particle in the circle is described by the vector $\underline{r}$ and the velocity of the particle by the vector $\underline{v}$.

Consider what is happening from the point of view of Newton's laws of motion, which must apply to all motions. Consider a time interval δt at the start of which the radius vector is $\underline{r}$ and the velocity is $\underline{v}$, and at the end of which the values are $\underline{r}'$ and $\underline{v}'$ respectively. This is shown in Fig. 2.5(b). Consider now the change in the velocity only; this is shown in Fig. 2.5(c) where it can be seen that the change $\underline{v}$ to v' requires the additional velocity $\delta\underline{v}$. To find the acceleration $\underline{a}$ undergone by the mass m, one simply has to find

$$\frac{d\underline{v}}{dt} = \underset{\delta t \to o}{Lt} \frac{\delta \underline{v}}{\delta t} .$$

As $\delta t \to o$, $\delta \underline{v} \to o$ and the angle ϕ (Fig. 2.5c) $\to \pi/2$ or 90°. When this is so, $d\underline{v}$, and hence $\underline{a}$, is directed always at right angles to $\underline{v}$ and towards the centre. Thinking in vector terms has simplified the argument leading to this result.

To find magnitudes, which are scalar quantities, one can argue in scalar terms. Inspection of Fig. 2.5(c) enables one to write

$$\frac{\delta \underline{v}}{\underline{v}} = \theta \tag{2.5}$$

Similarly, from Fig. 2.5(b),

$$\frac{AA'}{\underline{r}} = \frac{v\delta t}{r} = \theta . \tag{2.6}$$

From eqns (2.5) and (2.6), it follows that

$$\frac{\delta v}{v} = \frac{v\delta t}{r} ,$$

whence

$$a = \frac{dv}{dt} = \frac{\delta v}{\delta t} = \frac{v^2}{r} . \tag{2.7}$$

This result, that the acceleration towards the centre is (v^2/r), is well known. It is often useful to express it in angular terms; if ω is the angular speed of m about the centre of the circle, then $\theta = \omega\delta t$ and substitution in eqn (2.6) gives at once

$$v = \omega r. \tag{2.8}$$

For eqn (2.7) one can therefore write that

$$a = \frac{v^2}{r} = \omega^2 r. \tag{2.9}$$

Newton's laws of motion state explicitly that a force is necessary to produce a change in motion (from uniform straight-line motion or from rest), and then go on to indicate what the magnitude of the force must be in terms of the rate of change of momentum. For constant mass m, these statements are summarized by eqn (1.2) relating force $\underline{F}$ and acceleration $\underline{a}$, i.e., by

$$\underline{F} = m\underline{a}.$$

Therefore, if $\underline{a}$ is directed towards the centre, so too is $\underline{F}$. Thus, if a mass is moving at uniform speed round a circle, it is always subject to a force directed towards the centre of the circle. This force is known as the centripetal force, and it must be present if circular motion is to be achieved; from eqns (1.2), (2.7) and (2.9) its magnitude is $(mv^2)/r$ or $mr\omega^2$.

All children, who have whirled a weight on the end of a string, are aware that forces are necessary to achieve circular motion of the weight and that these forces act along the line of the string. But the force of which they are perhaps most aware is the reaction of the weight to their holding force: the outward pull of the weight on the string. Reference to D'Alembert's principle (Section 1.3) makes it clear that this is an inertial reaction, which has all the attributes of a force other than the rather important ability to produce a change in linear motion. It is commonly given the name of centrifugal force, but it is only a force in the sense that any other inertial reaction is a force. Certainly, if one is moving in a circle oneself, one can be acutely aware of its presence; but D'Alembert's principle should lead one to expect this awareness.

Eqn (2.8) related the scalar magnitudes of the vector quantities $\underline{v}$,$\underline{\omega}$ and $\underline{r}$ for the specific case of motion in a circle. Consider now the general case of angular velocity about a point, in three dimensions, where one wishes to find a general relation between these vector quantities. Let the point be P in Fig. 2.6. The direction of the angular velocity vector $\underline{\omega}$ will be that of the vector product $\underline{r} \times \underline{v}$; i.e. $\underline{\omega}$ will be at right angles to both

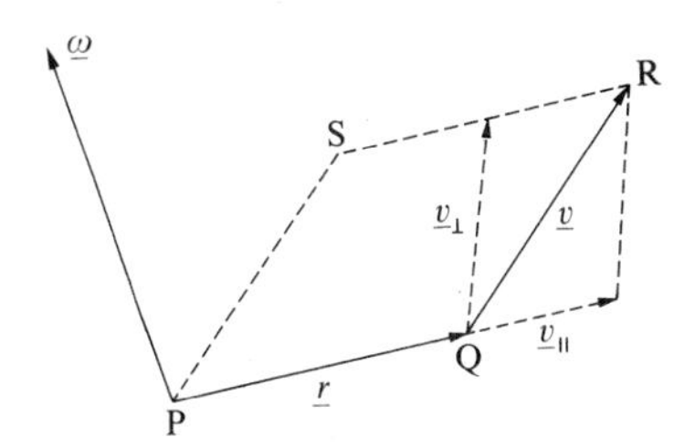

FIG. 2.6. To illustrate the angular velocity about a point P of a moving point Q.

$\underline{r}$ and $\underline{v}$. The next question is the magnitude of $\underline{\omega}$. Now the magnitude of $\underline{r}$ and $\underline{v}$ is the size of the area PQRS on Fig. 2.6, i.e., is $rv_\perp$, so

$$|\underline{r} \times \underline{v}| = rv_\perp . \qquad (2.10)$$

Eqn (2.8) showed from scalar considerations that

$$\omega = \frac{v_\perp}{r} . \qquad (2.11)$$

$v_\perp$ appears in this equation because, in the discussion which led to eqn (2.8), one was only concerned with $\underline{v}$ perpendicular to $\underline{r}$.

Substituting in eqn (2.10) the value of $v_\perp$ given by eqn (2.11), it follows that

$$|\ \underline{r} \times \underline{v}\ | = r^2\omega,$$

or, better, as the direction of ω is that of $\underline{r} \times \underline{v}$,

$$\underline{r} \times \underline{v} = r^2\underline{\omega}. \qquad (2.12)$$

Eqn (2.12) thus gives both the magnitude and direction of $\underline{\omega}$.

Going now from the general case to the specific case of motion in a circle, when r^2 is constant, eqn (2.12) can be differentiated with respect to time to give

$$\underline{r} \times \underline{a} = r^2\underline{\alpha}, \qquad (2.13)$$

$\underline{a}$ being the linear and $\underline{\alpha}$ the angular acceleration. In going from

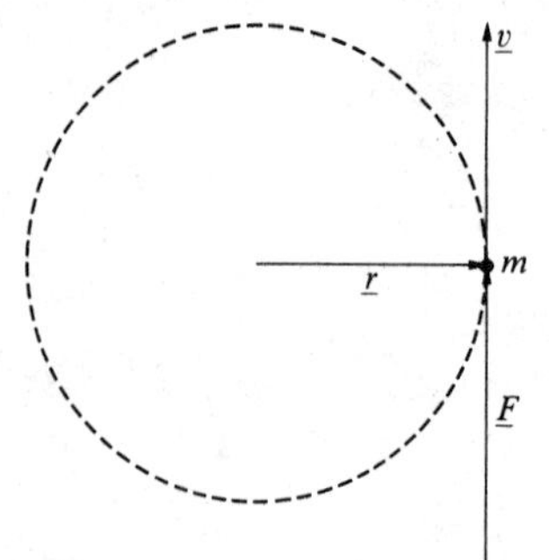

FIG. 2.7. To illustrate the discussion of the effect of a force on circular motion.

eqn (2.12) to eqn (2.13), one notes that $\dot{\underline{r}} \times \underline{v} = 0$ as $\dot{\underline{r}}$ is parallel to $\underline{v}$.

Consider now this circular motion when the speed round the circumference is allowed to change. Consider the situation illustrated by Fig. 2.7, where a force $\underline{F}$, collinear with $\underline{v}$, is acting on the mass m. Consider now only the motion perpendicular to $\underline{r}$ at the instant shown, when, if $\underline{m}$ is constant, Newton's laws, as expressed by eqn (1.2) state

$$\underline{F} = m\underline{a} \tag{2.14}$$

where $\underline{a}$ represents the component of total acceleration in the direction of $\underline{F}$ and $\underline{v}$. ($\underline{F}$ is of course not the total force acting on m; there must be a component due to the centripetal force as well). This $\underline{a}$ represents the instantaneous rate of change of the speed round the circumference. Taking now the vector product of $\underline{r}$ with both sides of eqn (2.14), one obtains

$$\underline{r} \times \underline{F} = m(\underline{r} \times \underline{a})$$

which, from eqns (2.3) and (2.13), gives

$$\underline{T} = mr^2\underline{\alpha}. \tag{2.15}$$

The quantity mr^2, which in eqn (2.15) relates the torque to the angular acceleration, is an important quantity known as the moment of inertia of the mass m about the centre of its circular motion. Any rotating mass can be described as having a moment of inertia about its axis of rotation; the property of moment of inertia is of sufficient importance in the theory of rotary motion to deserve fuller discussion.

2.4. MOMENT OF INERTIA

Consider some large massive object, as shown in Fig. 2.8, which is rotating about an axis through the point O and perpendicular

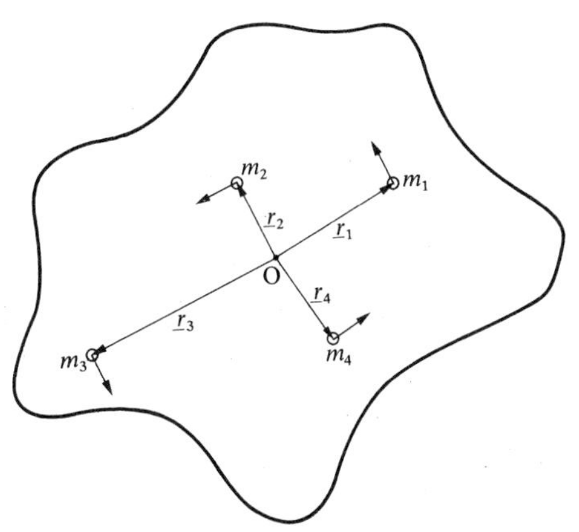

FIG. 2.8. A large body, rotating about an axis through O perpendicular to the plane of the diagram, can be looked on as a large number of small masses, m_n, moving in circles of radii r_n about O.

to the plane of the diagram. Imagine the object to be made from a large number of small (in effect, point) masses m_1, m_2, m_3 ---- m_n ---, distances r_1, r_2, r_3 ---- r_n --- from O. If the object is accelerating in its motion with angular acceleration $\underline{\alpha}$, then each mass, m_1, m_2, etc., is also accelerating at the rate $\underline{\alpha}$ about

O. So, for any one small mass, say for m_n, eqn (2.15) takes the the form

$$\underline{T}_n = m_n r_n^2 \underline{\alpha}, \tag{2.16}$$

where $\underline{T}_n$ is the torque required to accelerate m_n. The total torque $\underline{T}$ required to accelerate the whole of the object can be found by adding together all the torques required to accelerate all the small masses, i.e., from eqn (2.16),

$$\underline{T} = \Sigma \underline{T}_n = \Sigma m_n r_n^2 \underline{\alpha} = (\Sigma m_n r_n^2) \underline{\alpha}, \tag{2.17}$$

where the summation is taken over all the possible values of n. The quantity $(\Sigma m_n r_n^2)$ is known as the moment of inertia I of the object about this axis through O, so eqn (2.17) can be rewritten as

$$\underline{T} = I\underline{\alpha}. \tag{2.18}$$

Comparison of eqn (2.18) with eqn (1.2) makes it clear that, when torques are substituted for forces and angular accelerations for linear accelerations, moment of inertia becomes for angular motion what mass is for linear motion. One very important difference though is this. When stating a moment of inertia, it is essential to specify the axis of rotation, as the moment of inertia is not a scalar constant of the body concerned in the way that mass is.

The calculation of I from the expression

$$I = \Sigma_{\text{all } n} (m_n r_n^2) \tag{2.19}$$

depends in its complexity on the shape of the object and, for simple shapes, can reduce to a simple integration.

The analogy between I and mass can be further seen by considering the kinetic energy associated with rotational motion. Let the object of Fig. 2.8 be rotating with angular speed ω, and let the speeds of each of the masses m_1, m_2, etc., be v_1, v_2, etc.. The kinetic energy of the mass m_n will be given by

$$T_n = \tfrac{1}{2} m_n v_n^2.$$

Eqn (2.8) gives

$$v_n = r_n \omega,$$

$$\therefore \quad T_n = \tfrac{1}{2} m_n r_n^2 \omega^2.$$

The total kinetic energy T of the body is therefore

$$T = \Sigma_{\text{all } n} T_n = \Sigma_{\text{all } n} \tfrac{1}{2} m_n r_n^2 \omega^2 = \tfrac{1}{2} (\Sigma_{\text{all } n} m_n r_n^2) \omega^2,$$

i.e., from eqn (2.19),

$$T = \tfrac{1}{2} I \omega^2. \tag{2.20}$$

The obvious comparison between eqns (2.20) and (1.19) emphasizes the analogy between moment of inertia and mass. The kinetic energy of eqn (2.20) is just the same in nature as that of eqn (1.19); it was obtained by simply summing up a lot of equations of the type of eqn (1.19). In any event, kinetic energy is a scalar quantity, depending on speed rather than on velocity, so the fact that the motion is non-linear need not affect it.

2.5. TORQUE AND WORK

Suppose a rotating body, of moment of inertia I and angular velocity $\underline{\omega}$, is slowed down by a torque $\underline{T}$ with uniform angular acceleration $-\underline{\alpha}$ and brought to a halt: suppose too that the angle moved through by the body while being slowed down is θ.

Using $|\underline{T}|$ for the magnitude of $\underline{T}$ to avoid confusion with kinetic energy, eqn (2.18) for this case gives

$$|\underline{T}| = -I\alpha. \tag{2.21}$$

The kinematic equation for angular motion analogous to eqn (1.9) gives

$$\theta = -\frac{\omega^2}{2\alpha} \,. \tag{2.22}$$

From eqns (2.21) and (2.22), it follows that

$$|\underline{T}|\theta = \tfrac{1}{2} I\omega^2,$$

so that the product of torque times angle moved through by torque has the dimensions of energy; it is in fact a measure of the work done by the torque in slowing the body down. Thus torque × angle moved through by torque = work done by torque.

2.6. ANGULAR MOMENTUM

The analogy between moment of inertia and mass may profitably be extended beyond energy and the equation of motion so far used, i.e., eqn (2.18). Starting with this last equation, one notes

$$\underline{T} = I\underline{\alpha} = I\frac{d\underline{\omega}}{dt} = \frac{d}{dt}(I\underline{\omega}) \tag{2.23}$$

provided that I does not depend on $\underline{\omega}$. By analogy between eqns (2.23) and (1.1), the quantity $I\underline{\omega}$ can be defined as the angular momentum, $\underline{L}$; eqn (2.23) then states that torque = rate of change of angular momentum. It is now relevant to see the relationship between angular momentum and linear momentum. Consider the uniform motion of a mass m around a circle, as in Fig. 2.5(a). First, one can substitute for $\underline{\omega}$ using eqn (2.12), so that

$$\underline{L} = I\underline{\omega} = I\frac{\underline{r} \times \underline{v}}{r^2} \,. \tag{2.24}$$

For this simple case, $I = mr^2$, so that

$$\underline{L} = mr^2 \frac{\underline{r} \times \underline{v}}{r^2} = \underline{r} \times (m\underline{v}), \tag{2.25}$$

whence

$$\underline{L} = \underline{r} \times \underline{p}. \tag{2.26}$$

This is the result relating angular to linear momentum. Comparison of this with the expression for a torque or the moment of a force, eqn (2.3), shows one the reason for an alternative name often given to angular momentum, namely moment of momentum. One could in fact have used eqn (2.26) as a definition and argued from there, but there is no advantage in doing so. The first important result is anyway quite clear; angular momentum is a vector quantity of magnitude $I\omega$.

If an axis of rotation is clearly defined, then the angular momentum about that axis is also clearly defined and is a vector quantity lying along that axis in the same direction as the angular velocity vector.

Angular momentum about any particular axis is conserved in the same way as linear momentum. If one considers the problem of two bodies, rotating about the same axis, brought into collision, then an argument similar to that used in Section 1.8 for linear momentum will soon show this to be true. An example of a practical consequence of this conservation is provided by a skater who is rotating rapidly on the ice; if he or she wishes to speed up the rotation, his or her arms will be brought close to the body, so reducing 'I'; for $I\omega$ to remain constant, ω then increases. Slowing down of rotation is achieved by the reverse process of stretching the arms out.

2.7. SIMPLE GYROSCOPIC PRECESSION

Simple gyroscopic precession is a form of motion which can be described and understood by reference to angular momentum as a vector quantity. This precession can be illustrated by a form of child's toy consisting primarily of a small heavy flywheel and

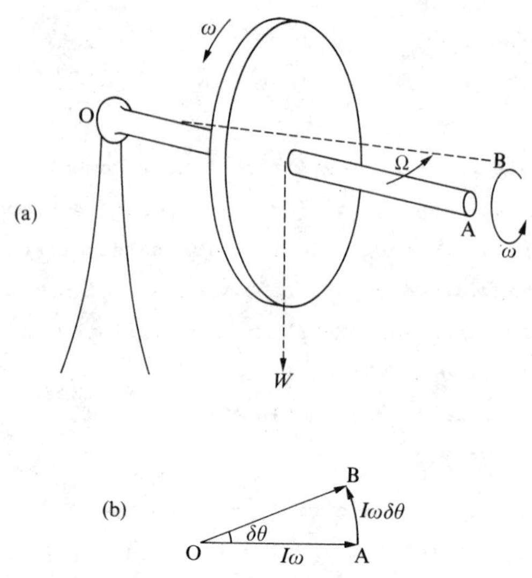

FIG. 2.9. To illustrate gyroscopic precession; (a) shows a simple toy gyroscope precessing around a vertical axis through a support O, and (b) shows the relevant angular momentum vector diagram viewed from above.

axle which can be put into rapid rotation; one end of the axle can in effect be rested on a support and the whole is held horizontally and released. On release, the gyroscope, i.e. the flywheel and axle, rotate around the support as shown in Fig. 2.9(a).

The explanation of this motion is now really quite simple.With reference to the figure, consider the flywheel to be initially in the position indicated as OA. In this position, the flywheel both has angular momentum $I\underline{\omega}$ along OA and is acted on by a torque, $\underline{G}$, say, given by the weight of the flywheel times the distance of its centre of gravity from O. The direction of this torque $\underline{G}$ will be given by the right-handed corkscrew rule.

Consider now what happens. The gyroscope precesses, i.e., rotates around the support, and in so doing moves round O with some constant angular velocity $\underline{\Omega}$. If it starts at position OA, a small time δt later it will have precessed through an angle $\delta\theta$

to reach position OB, as shown in Fig. 2.9. The change in angular momentum will be that shown in Fig. 2.9(b) as the line AB, having magnitude $I\omega\delta\theta$ and direction that of the torque $\underline{G}$. The rate of change of angular momentum in that same direction is thus $I\omega(\delta\theta/\delta t)$, i.e., $I\omega\Omega$. Considering magnitudes only, the equation of motion in the form 'torque equals rate of change of angular momentum', i.e. eqn (2.23), can now be applied to give

$$G = I\omega\Omega. \tag{2.27}$$

Eqn (2.27) can very often be used to calculate precession velocities. It should be noted though that the angular momentum about the axis of precession has been neglected; if this angular momentum is not very much less than $I\omega$, the motion can easily turn out to be more complicated. However, if the motion is just a simple precession, it is interesting to note that the faster the flywheel of the gyroscope rotates, the slower it precesses as, for a given torque, the product ω (rotation speed) $\times$ Ω (precession speed) is a constant. Similarly, as it slows down, its precession speeds up, as any child who has played with such a toy will have noted.

Gyroscopic precession is a phenomenon which is not restricted to large-scale events. An electron possesses angular momentum (spin); if a couple is exerted on it by a magnetic field acting on the magnetic moment associated with the spin, it will precess. This phenomenon is known as the Larmor precession and is important in the history of atomic theory.

2.8. THEOREMS OF PARALLEL AND PERPENDICULAR AXES

The fact has already been mentioned that moment of inertia, although analogous to mass, is not a constant of the body concerned in the way that mass is. The moment of inertia of a rotating body depends on the axis about which the rotation is taking place. The value of the summation for I, given by eqn

(2.19), namely

$$I = \sum_{\text{all } n} (m_n r_n^2), \tag{2.28}$$

must clearly depend on the position of the axis, which fixes the values of r_n. The ease of calculating this summation depends on the shape of the body and the position of the axis: this can be illustrated with the help of a simple example. Consider a flat uniform circular disc, as shown in Fig. 2.10. Let this disc has mass M and radius R, and consider the calculation of the moment of inertia about an axis through the centre of mass O perpendicular to the plane of the disc. This axis is the easiest one about which to calculate the moment of inertia because, as will be shown, the summation turns into a simple integral. The calculation now follows.

As the disc is uniform, let the mass per unit area be ρ, so that

$$\rho = M/\pi R^2. \tag{2.29}$$

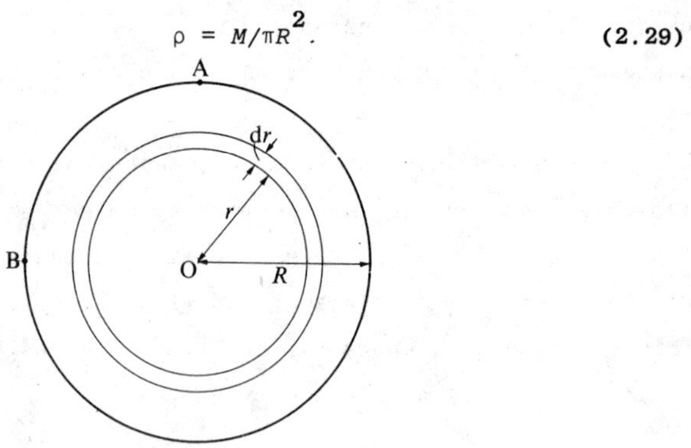

FIG. 2.10. Flat circular disc, centre O, of radius R.

Now let the m_ns of eqn (2.28) be annuli, as depicted in Fig. 2.10, of inner radius r and outer radius $r + \mathrm{d}r$, where r lies anywhere between 0 and R, and $\mathrm{d}r \ll r$. The value of each m_n is then ρ times its area, i.e. ρ times $2\pi r\mathrm{d}r$, and the value of each r_n is r, so one can rewrite eqn (2.28) as

$$I = \Sigma(\rho . 2\pi r \mathrm{d}r . r^2), \qquad (2.30)$$

where the summation is over all r from 0 to R. As dr is infinitesimally small, this can be rewritten as an integral, so eqn (2.30) becomes

$$I = \int_0^R 2\rho\pi r^3 \mathrm{d}r,$$

which, on integrating, gives

$$I = \tfrac{1}{2}\rho\pi R^4,$$

which in turn, using eqn (2.29), gives

$$I = \tfrac{1}{2}MR^2. \qquad (2.31)$$

This calculation has been easy because the choice of axis and the shape of the disc have made it so; it has been easy both to divide the 'm_ns' up into a form which is easily expressible as a function of r and to put convenient and useful limits on to the integration. Suppose, however, that one wishes to calculate the moment of inertia of the disc about any other axis than the one chosen. A little consideration of the problem will soon show that it may not be as easy to convert the summation into an easily carried out integration. It may prove difficult to find suitable small elements and to fit appropriate limits. There are, however, two useful theorems which can be applied for certain other axes. These are the theorems of parallel and perpendicular axes, and these will now be discussed in turn.

2.8.1. *Theorem of parallel axes*

This theorem is applicable to all shapes of rotating objects. It states quite simply:

The moment of inertia of a body about any axis (I_A) is equal to its moment of inertia about a parallel axis through its centre of mass (I_c) plus the mass of the body (m) times the square of the distance between the axes (d), i.e.,

$$I_A = I_c + md^2. \tag{2.32}$$

The usefulness of this theorem can be seen by reference to the point A on Fig. 2.10. Suppose that one wishes to calculate the moment of inertia of the disc about an axis through A and perpendicular to the plane of the disc. The moment of inertia about the appropriate parallel axis through the centre of mass O has already been seen to be easy to calculate, and is given by eqn (2.31); eqn (2.32) now allows one to write down immediately the desired moment of inertia I_A, giving

$$I_A = \tfrac{1}{2}MR^2 + MR^2,$$
$$= \frac{3}{2}MR^2.$$

It would in principle have been possible to have set up an integral which, on evaluation, would have given this result; however, such a procedure would have been both lengthy and clumsy, hence unprofitable when the theory of parallel axes can so easily be applied.

Proving the theorem to be true is relatively easy and is left to the reader as an exercise. While it is often satisfying to oneself to have proved that a theorem is valid, it is however not necessary to have done so in order to be able to put the theorem to use; this is particularly true of such a well worked-over case as this one.

2.8.2. *Theorem of perpendicular axes*

This theorem is much more restricted in its applicability than the theorem of parallel axes because it requires the body to be a plane object, stictly speaking two-dimensional only, although the theorem will be useful if the thickness of the plane is negligible compared with the size in the other two dimensions.

The use of the theorem in real cases is thus bound to be an approximation as no real body is infinitely thin, but approximate results are often all that is needed. The theorem may be stated as follows:

The moment of inertia of any uniform plane body about any axis perpendicular to the plane of the body (I_z) is equal to the sum of the moments of inertia of the body about any two perpendicular axes in the plane of the body (I_x,I_y) which intersect on the first axis,
i.e.,

$$I_z = I_x + I_y. \tag{2.33}$$

Again, this theorem is very easy to prove and the proof is left as an exercise. For an illustration of its use, the body shown in Fig. 2.10 can again provide an example, if one assumes it now to be very thin. Suppose that one wishes to calculate the moment of inertia of the disc about any diameter, I_{D}. As it is a uniform circular disc, the moments of inertia about all diameters will be the same. Eqn (2.31) states that the moment of inertia about an axis perpendicular to the plane of the body through O, I_{o}, is equal to $\frac{1}{2}MR^2$. If OA and OB represent the directions of two mutually perpendicular axes intersecting at O, then, using eqn (2.33), the theorem of perpendicular axes gives

$$I_{\mathrm{o}} = I_{\mathrm{oA}} + I_{\mathrm{oB}}.$$

But

$$I_{\mathrm{oA}} = I_{\mathrm{oB}} = I_{\mathrm{D}},$$

therefore

$$\tfrac{1}{2}MR^2 = 2I_{\mathrm{D}},$$

giving

$$I_{\mathrm{D}} = \tfrac{1}{4}MR^2. \tag{2.34}$$

Use of the theorem is again simpler than setting up the appropriate integral and evaluating it. It should be noted that, in mathematical jargon, a plane body of this type is often referred to as a lamina.

The theorems of parallel and perpendicular axes are of course more generally applicable than to the example given of a circular disc; this shape was chosen solely to provide a simple illustration.

2.9. RADIUS OF GYRATION

If a body of mass M has a moment of inertia I about some given axis, then its radius of gyration k about that same axis is defined by the equation

$$I = Mk^2.$$

It is clearly often not necessary to use the term 'radius of gyration' as well as 'moment of inertia'; however, it may sometimes be felt to be useful and it does occur in the literature.

2.10. MORE ABOUT MOMENT OF INERTIA

So far, the quantity moment of inertia I has appeared in two vector equations, namely eqn (2.18), i.e.,

$$\underline{T} = I\underline{\alpha}, \tag{2.35}$$

where it relates the torque $\underline{T}$ to the angular acceleration $\underline{\alpha}$, and the first part of eqn (2.24), i.e.,

$$\underline{L} = I\underline{\omega}, \tag{2.36}$$

where it relates angular momentum $\underline{L}$ to angular velocity $\underline{\omega}$. In the discussion so far, it has been implied that $\underline{T}$ and $\underline{\alpha}$ have the same direction ahd that $\underline{L}$ and $\underline{\omega}$ have the same direction, and in the examples given it has been true that all these pairs of directions have been the same. The quantity I has been treated as though it were a scalar quantity, but it has been emphasized that it is not a constant of the body concerned in the way that mass is, an emphasis which contains the implication that it might have non-scalar properties. It has also been made clear that it depends on the axis about which it is taken; eqn (2.32), i.e., the theorem of parallel axes, has however indicated that, for any one set of parallel axes, the moment of inertia takes a minimum value for rotation about the axis which passes through the centre of mass. There is, though, an infinite number of axes, all with different directions, which pass through the centre of mass; for the case of the flat disc, eqns (2.31) and (2.34) gave the moments of inertia about two of these and they were not the same. This shows that there is something directional about the moment of inertia of a body; even when the chosen axis passes through the centre of mass, the direction of the axis can affect the value of I.

Clearly then I is not a simple scalar quantity. An analysis of what it is shows that, in mathematical terms, it is a tensor (strictly, a second-order tensor) quantity. Such tensor quantities turn up from time to time in physics; the study of solid single crystals, for example, is a study of substances whose simple properties (such as refractive index, resistivity, thermal conductivity etc.) can depend markedly on direction in the crystal; the description of such directional properties can again usefully introduce the tensor quantity. Section 2.11 goes into this, as far as I is concerned, in a little more detail. However, for many purposes, it is sufficient to be aware of the fact that, in certain circumstances, $\underline{T}$ and $\underline{\alpha}$ and also $\underline{L}$ and $\underline{\omega}$ may not be collinear, the quantity I in eqns (2.35) and (2.36) having non-scalar and direction-changing properties.

2.11. THE ELLIPSOID OF INERTIA

Take any point in any body and treat it as the origin of

cartesian coordinates. Let the body rotate with angular velocity $\underline{\omega}$ about any axis through this point, and let it have a moment of inertia $\underline{I}$ about this axis. Consider a vector $\underline{r}$ in the direction of $\underline{\omega}$ whose magnitude is given by

$$r = I^{-\frac{1}{2}}. \qquad (2.37)$$

If all possible axes through this point are now considered, the various vectors $\underline{r}$ will become radius vectors to a surface. It turns out that, whatever the shape of the body or whatever point in space is chosen, this surface is an ellipsoid; this result is worth knowing. The ellipsoid so formed is known as the ellipsoid of inertia; it will have principal axes, x, y, z, such that the equation of the surface becomes

$$I_x x^2 + I_y y^2 + I_z z^2 = 1, \qquad (2.38)$$

where I_x, I_y, I_z are the moments of inertia of the body about the x, y, and z axes, and are known as the principal moments of inertia.

The simple flat circular disc, already discussed in earlier sections, provides a ready example; if the point to be considered is the centre of mass and if the disc lies in the xy plane of the coordinates, then eqns (2.34) and (2.31) tell one that

$$I_x = I_y = \tfrac{1}{2} I_z. \qquad (2.39)$$

Simple considerations of symmetry show that eqn (2.39) may be taken to describe principal moments of inertia; thus in this case the ellipsoid of inertia, defined by eqn (2.38), is an ellipsoid of revolution about the z axis, flattened in the z direction as, from eqn (2.37), the radius vector in that direction is $2^{-\frac{1}{2}}$ times the radius vector anywhere in the xy plane. In the general case, it may be noted that the existence of the ellipsoid of inertia simplifies the calculation of I about any axis once the principal moments of inertia are known.

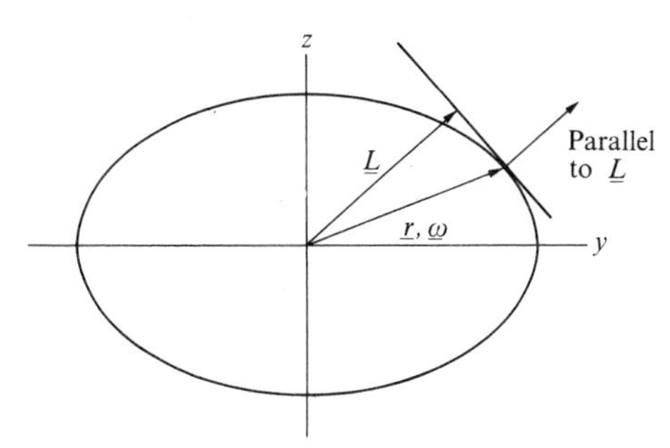

'IG. 2.11. To illustrate the relationship between $\underline{L}$,$\underline{\omega}$, and the ɘllipsoid of inertia defined by the radius vector $\underline{r}$

The ellipsoid of inertia may also be used as an indicator of :he relative directions of $\underline{L}$ and $\underline{\omega}$. Fig. 2.11 shows how this is lone. Suppose that the body is rotating with angular velocity) about an axis through a point about which the ellipsoid of .nertia has been drawn. (For convenience in drawing the diagram, :his axis has been taken to be in the plane of the diagram, the 'z plane). The radius vector $\underline{r}$ in the direction of $\underline{\omega}$ meets the ellipsoid at point P. The angular momentum $\underline{L}$ is then parallel to he normal to the ellipsoid at P and passes through the point ıbout which the rotation is taking place. The theory behind this ıethod of relating the directions of $\underline{L}$ and $\underline{\omega}$ is beyond the scope of this book, but the interested student may care to consult a ıore advanced text on this point, or other unproved points on this opic, but be advised not to waste too much time on such a follow-ıp at this stage. However, it is important to note that if the lirection of $\underline{\omega}$ is along one of the principal axes of the ellipsoid of inertia, then $\underline{\omega}$ and $\underline{L}$ lie in the same direction. 'he relative directions of $\underline{T}$ and $\underline{\alpha}$ can be similarly found as I .cts both as a scaling factor and a direction changer in the same 'ay in both of the eqns (2.35) and (2.36).

The existence of an ellipsoid of inertia with principal ıxes, which will be axes for rotations about which $\underline{L}$ and $\underline{\omega}$ will e parallel, does not depend on the shape or symmetry of the body in any way. However, the symmetry of a rotating body can often pin down at once the location of a principal axis of the ellipsoid of inertia, a point which was used earlier for the flat circular disc. Thus if a body shows at least a simple two-fold symmetry about the axis of rotation, then that axis is such a principal axis and $\underline{L}$ and $\underline{\omega}$ will both lie along it. Such simple two-fold symmetry means that a rotation of π (180°) leaves the body identical, as far as the location of mass is concerned, before and after the rotation; in crystallographic terms, this is a diad axis. Axes of higher symmetry are also principal axes, and when the symmetry is higher one can expect the ellipsoid of inertia to be an ellipsoid of revolution about that axis of symmetry. If the body shows no obvious symmetry, then clearly the location of principal axes is not self-evident.

2.12. AN EXAMPLE OF AN ASYMMETRICALLY ROTATING BODY

The discussion of this section is intended solely to illustrate the consistency of the arguments of preceding sections.

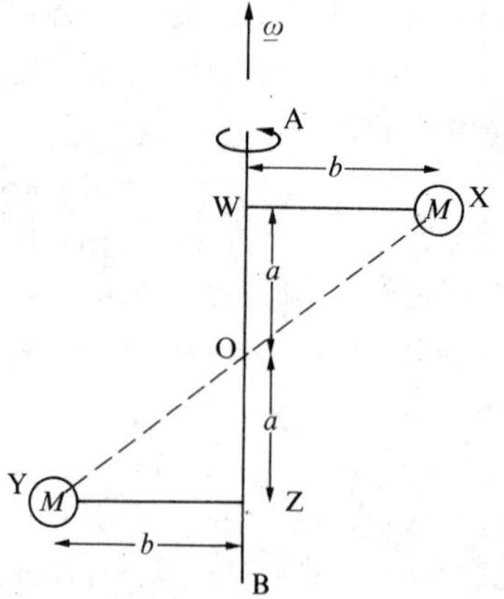

FIG. 2.12. A body rotating about an axis which is not an axis of symmetry of the body.

Consider the body illustrated in Fig. 2.12. This consists of two masses, each of magnitude M, joined rigidly by a system of light rods,

WX, YZ and AB; by light rods are meant rods whose masses may be neglected relative to M for the purpose of the discussion. The masses and rods all lie in one plane, and the distances are as marked in the diagram. Let the whole body now rotate about the axis AB with angular velocity $\underline{\omega}$ in the direction shown.

Without considering moments of inertia, but only considering the centrifugal forces which the masses M will exert on the rod AB, it is immediately apparent that the body will at any instant be subjected to a torque about O. From the discussion in Section 2.3, the magnitude of each force will be $Mb\omega^2$. These, being centrifugal, are forces of inertial reaction and provide a torque of inertial reaction about O whose magnitude will be $2a(Mb\omega^2)$ and whose direction is perpendicular to the plane of the diagram and away from the viewer. Therefore to keep in its place the axis about which the body is rotating, an equal and opposite torque $\underline{T}$ must be applied, whose magnitude is given by

$$T = 2Mab\omega^2, \qquad (2.40)$$

but whose direction is now perpendicular to the plane of the diagram and towards the viewer. The bearings holding the rotating rod AB, wherever they are located, must provide this torque.

Torques of this sort can be important to engineers in the design of rotating machinery; the bearings and the framework holding the bearings must be strong enough to take the relevant forces. The argument given so far is sufficient for the purpose of calculating the torque.

However, to show that the discussions in earlier sections are consistent, it is desirable to see how the presence of this torque is to be reconciled with the equation

$$\underline{T} = \frac{d\underline{L}}{dt} \qquad (2.41)$$

which is a restatement of eqn (2.23), stating that the torque equals the rate of change of angular momentum, $\underline{L}$.

In order to find the angular momentum $\underline{L}$, it is convenient first to construct the ellipsoid of inertia for rotations about O, with the help of eqn (2.37). If the masses M are treated as 'point' masses (i.e. of negligible size) and if one still neglects the various rods, then the moment of inertia for rotation about the axis XOY is zero (or negligible) and, from eqn (2.37), the magnitude of the radius vector of the ellipsoid of inertia in this direction is infinite (or at least very large). One soon finds that the ellipsoid of inertia in this case can be looked on as in effect a right circular cylinder with axis XOY and radius $\{2M(a^2 + b^2)\}^{-\frac{1}{2}}$. A cross-section of this cylinder in the plane of the body is shown as Fig. 2.13. Application now of the construction

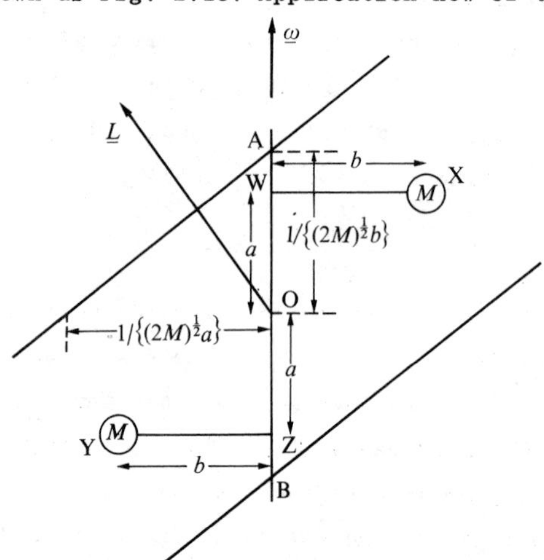

FIG. 2.13. The central part of the ellipsoid of inertia of the body shown in Fig. 2.12; it is approximately a circular cylinder about the line XOY as axis; the construction given in Fig. 2.10 shows that the angular momentum vector for the body, when rotating about any axis through O, will always be perpendicular to the line XOY.

shown in Fig. 2.11 shows at once that the total angular momentum of the rotating body lies perpendicular to XOY in the plane of the body, as shown in Fig. 2.13. (Note how the asymmetry about the axis of rotation has led to non-collinearity of $\underline{L}$ and $\underline{\omega}$).

The magnitude of $\underline{L}$ can be calculated as follows. Eqn (2.25), considering that there are two masses, gives

$$\underline{L} = 2(\underline{r} \times M\underline{v}),$$

which produces in this case

$$L = 2(a^2 + b^2)^{\frac{1}{2}} Mb\omega. \tag{2.42}$$

It may be noted that the alternative name for angular momentum, namely moment of momentum, provides a useful hint in deciding how to calculate $\underline{L}$.

Now refer the problem to cartesian axes x, y, and z, centred at O. Let the rotation be about the z axis and consider the instant when the body lies in the yz plane, as shown in Fig. 2.14.

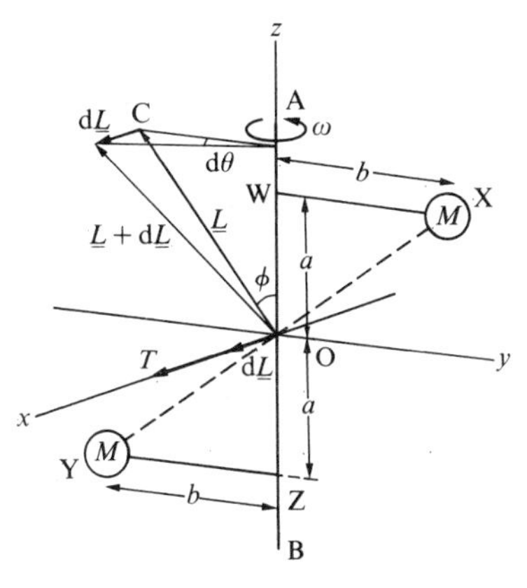

FIG. 2.14. The asymmetric rotating body of Fig. 2.11 with torque and angular momentum vectors added to illustrate the argument relating the torque to the velocity of rotation.

At an instant of time dt later, the body will have rotated through an angle dθ about z, and the angular momentum $\underline{L}$ will have changed by an amount d$\underline{L}$ as shown on the figure. d$\underline{L}$ is about O, so is located there, but can be redrawn in the diagram to be at C for discussion purposes. Consider now the triangle based at C of apex angle dθ; as dL and dθ are small, one can write

$$\frac{dL}{L \sin \phi} = d\theta,$$

therefore

$$\frac{dL}{dt} = L \sin \phi \frac{d\theta}{dt}. \tag{2.43}$$

Now $\sin \phi = a/(a^2 + b^2)^{\frac{1}{2}}$, $d\theta/dt = \omega$ and L is given by eqn (2.42); making these substitutions in the right-hand side of eqn (2.43) and using eqn (2.41), it follows that

$$T = \frac{dL}{dt} = 2Mab\omega^2. \tag{2.44}$$

One sees therefore at once that eqns (2.44) and (2.40) are identical, showing the consistency of the arguments. One also notes, from Fig. 2.14 as well as from eqn (2.41), that $\underline{T}$ and $d\underline{L}$ have the same direction.

An interesting aspect of this is that torque and rate of change of angular momentum are neither the cause nor the effect of each other, merely necessary accompaniments. One instinctively thinks of a force producing a change, because one applies a force in real life for this purpose. In this case though, similar reasoning would have made the rate of change of angular momentum the cause of the torque. But there is nothing in the laws of mechanics to distinguish cause and effect in this sort of way; it is only our own common-sense that allows such a distinction, and in real life, this distinction may have to be made.

Had the symmetry of the body been as shown in Fig. 2.15 (which is in effect the same body rotating about a different axis) clearly both $\underline{L}$ and $\underline{\omega}$ would have been in the same direction and there would have been no torque. This is an example of the two-fold symmetry discussed in Section 2.11. However, had the masses of Fig. 2.15 not been equal, and had there been a similar rotation about the centre of mass, it can easily be shown that $\underline{L}$ and $\underline{\omega}$ would again be collinear. So symmetry is not the only

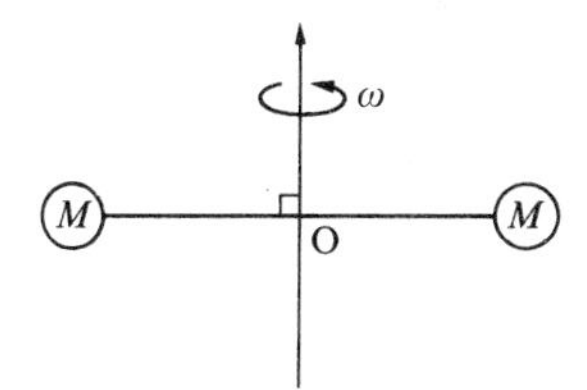

FIG.2.15. A symmetrical body, consisting of two masses M separated by a light rod rotating about an axis perpendicular to the rod and through the centre of mass O. The symmetry arises from the symmetry of rotations about such an axis.

criterion for locating a principal axis of an ellipsoid of inertia.

This section has been illustrative; use of the ellipsoid of inertia for the problem described would be a case of taking a sledge-hammer to crack a nut. It does emphasize, though, that care should be taken when relating total angular momentum, as opposed to a component of angular momentum, to angular velocity.

2.13. THE CORIOLIS FORCE

Why do the winds blow clockwise round the centre of an anticyclone in the northern hemisphere, but anticlockwise in the southern hemisphere? A glib answer to this question is simply 'because of the Coriolis force', but one will be none the wiser for this answer unless one knows what the Coriolis force is. It is the purpose of this section to explain it and to point out that there is nothing mysterious about it; it is a natural consequence of the laws of motion, just as centrifugal force is.

The explanation comes most simply with the help of an example. Consider the circular platform of Fig. 2.16, which is rotating about its centre O with uniform angular velocity $\underline{\omega}$. Consider a man standing at point A on the radius OC as the platform rotates; a short time later, OC will have moved to OC′ and the man from A to A . As the man moves with the platform, he will experience the centripetal force from the platform on to his feet and also an outward centrifugal force on his centre of mass; this latter force is his own inertial reaction (see Sections 1.3 and 2.3).

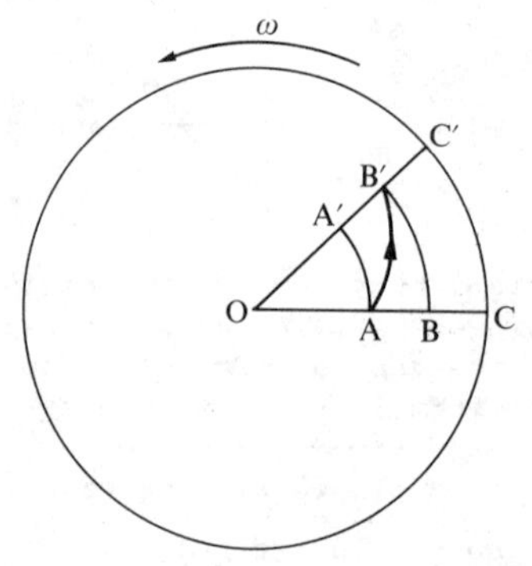

FIG. 2.16. To illustrate the discussion of the Coriolis force. The diagram shows from above a circular platform rotating with uniform angular speed ω. A man standing at point A moves directly outwards on a rotating radius, so reaching point B', while the radius moves from OC to OC'; his path in space is thus AB', while on the platform this would just seem to be AB.

These two forces are always directed towards or away from O. Provided that the man stays where he is on the platform, they will be the only forces experienced by him (apart, of course, from his weight, etc.). If, however, he moves on the platform, this may no longer be true. Let it be supposed that he decides to walk away from O along the rotating radius OC in order to reach the point B, and that he does this journey in the time that the radius takes to rotate from OC to OC′. His path in space will thus be the path AB′, although, relative to the rotating platform he has just moved directly away from O.

During the motion of the man from A to B′, the platform has continued to rotate at constant angular velocity $\underline{\omega}$. The moment of inertia I of the man about O (if m is the mass of the man and r his distance from O, $I = mr^2$) has increased (as r has increased) So the angular momentum $I\underline{\omega}$ has increased. To increase angular momentum, a torque has to be applied. As far as the man is concerned, a torque on him about O means a force F applied at right angles to the line OC along which he is walking; this force must be applied to his feet by the rotating platform as he moves. The equation of motion, eqn (2.23), equating torque to rate of change of angular momentum can be applied to find the magnitude of F; using scalar terms (a simplification possible because all vector quantities are either at right angles or parallel to each other), the equation becomes

$$T = rF = \frac{\mathrm{d}}{\mathrm{d}t}(mr^2\omega) = 2mr\omega\frac{\mathrm{d}r}{\mathrm{d}t},$$

so that

$$\mathrm{F} = 2m\omega\frac{\mathrm{d}r}{\mathrm{d}t},$$

$$= 2m\omega\dot{r}. \tag{2.45}$$

Thus, to keep the man on the line OC as it rotates and as he is moving outwards, the platform exerts on him a force $2m\omega\dot{r}$ directed at right angles to OC. From his point of view, he feels this force on his feet but experiences an equal and opposite force of inertial reaction; this means that, as he faces and walks outwards along the rotating line OC of Fig. 2.16, he feels this force pulling him over to his right. This force is the Coriolis force and, like centrifugal force, it is a force of inertial reaction. There is no specific name for the accelerating force F against which the Coriolis force reacts, but the associated acceleration, $2\omega\dot{r}$, is known as the Coriolis acceleration. In vector terms, the Coriolis force $\underline{F}_{\mathrm{Co}}$ is given by the equation

$$\underline{F}_{\mathrm{Co}} = -2m\underline{\omega} \times \dot{\underline{r}}, \tag{2.46}$$

the Coriolis acceleration being $2\underline{\omega} \times \dot{\underline{r}}$. It is important to note that the $\dot{\underline{r}}$ used in eqn (2.46) is measured with respect to the moving platform.

One example of the way atmospheric currents are affected by the Coriolis forces is illustrated by Fig. 2.17. Point A represents a point of low pressure in the earth's atmosphere in the northern hemisphere. Air moving in to fill this low-pressure region from the north along the line BA has an outward '$\dot{\underline{r}}$' as it is moving

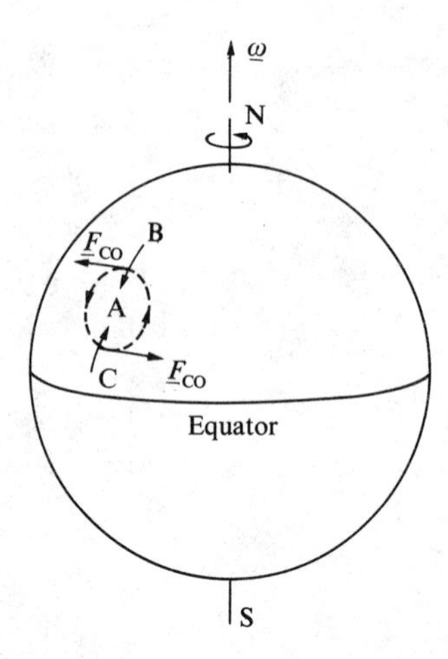

FIG. 2.17. To illustrate the effect of Coriolis forces, $\underline{F}_{Co}$ on the movement of the earth's atmosphere as the air moves in to the region A where the pressure is low.

away from the earth's axis of rotation; similarly, air from the south moving along CA has an inward '$\dot{\underline{r}}$'. The resulting Coriolis forces $\underline{F}_{Co}$ are shown, with their directions leading to a circulation of air around A as indicated by the arrows on the dashed circle. Around a similar low-pressure point in the southern hemisphere, the effect will produce a circulation in the opposite direction.

Apart from local regions of high or low pressure, one has also to take into account the Coriolis force on rising warm equatorial air; this leads to a general wind pattern which is clockwise in the northern hemisphere and anticlockwise in the southern.

2.14. INERTIAL AND NON-INERTIAL FRAMES OF REFERENCE

The term 'frame of reference' means the set of coordinates relative to which one is describing positions or movements. Very often, when a statement is made, a frame of reference is implicit but unmentioned. For example, if one states 'the earth moves in an orbit round the sun', one knows intuitively what one means;

one is attaching a reference frame of coordinates to the sun and stating that the earth moves in an orbit relative to this frame. Cosmologists of 2000 years ago could not agree on whether 'the earth moved round the sun' or 'the sun moved round the earth', but failed to see that the argument was partly only one of choice of coordinate frame. If one fixes a frame rigidly to the earth (a set of cartesian axes with the centre of the earth as origin, one axis through the poles and the others emerging at fixed locations on the equator, would be such a frame), then one can correctly state that the earth is stationary and the sun moves round it; intuitively, though, one would tend to be unhappy about the statement unless one incorporated into it the phrase 'relative to the earth'. If on the other hand one returns to fixing the reference frame to the sun, one can equally correctly state that the earth rotates on its axis and also moves in an orbit round the sun. Intuitively, one tends to be happier about this statement, partly because one learns about the solar system in childhood; had one not done so, one's intuitions might be different. One still remains unenquiring about it when learning that sunspot movements indicate rotation of the sun, accepting intuitively that same implicit reference frame without defining it.

Consider now the adjectives 'inertial' and 'non-inertial' when applied to reference frames. The meanings are simple: an inertial reference frame is one in which Newton's laws of motion are obeyed, and a non-inertial frame one in which they are not.

One can now see why one intuitively accepts one reference frame rather than the other. The cartesian axes fixed to the earth formed a non-inertial frame: all the distant stars move in it in non-linear paths without apparent forces contrary to Newton's first law, so one tends as a scientist instinctively to reject it. The frame fixed to the sun is instinctively accepted provided that one accepts that the sun's rotation takes place in this fixed frame, a frame centred at the sun but such that distant stars do not rotate about its origin. This intuition has guided one correctly, but is itself based on some insight into the laws of mechanics.

The discussion in Section 2.13 on Coriolis forces provides a good example of inertial and non-inertial reference frames. When discussing the circular platform of Fig. 2.16, the rotation of that platform was implicitly relative to an inertial reference frame. The result quoted as eqn (2.46) incorporated a vector quantity $\dot{\underline{r}}$ which was measured relative to a non-inertial frame, namely a frame fixed to the rotating platform in which the quantity $\dot{\underline{r}}$ remained in the same straight line. The Coriolis acceleration arose from variations in $\dot{\underline{r}}$ relative to the inertial frame. The forces of inertial reaction (see Sections 1.3 and 2.3 as well as Section 2.13) are experienced by observers when their own personal reference frame has ceased to be inertial, and to these observers they are real observable effects even if they do not enter into the equations describing their motion in an inertial reference frame.

The fact that some reference frames are inertial, and others not, is extremely fundamental to the description of dynamical phenomena provided by classical mechanics. It seems that, in order to provide a useful description, it is necessary to choose a certain type of reference frame; essentially, the type of reference frame chosen is the type in which the description proves to be the most simple.

The problem as to whether or not there is an absolute space in which one fixes inertial frames remains unsolved. Ernst Mach, whose definition of mass was quoted in Section 1.2.1, took the view that there was no absolute space and inertial frames were fixed by the presence of matter in the universe; in effect, the presence of the distant stars fixed the reference frame. But one cannot remove all the distant stars in order to put this hypothesis, often known as Mach's principle, to experimental test;

all one knows is that there appear to be such inertial reference frames, irrespective of their cause.

2.15. CONCLUSION

Rotational motions are so important that it is essential for the physicist, chemist or engineer to be able to deal with their theory; the concept of angular momentum is fundamental in much of quantum theory. Inertial and non-inertial reference frames are important in the development of the theory of relativity. This chapter has provided an introduction to all of these topics and at the same time one hopes that it will have encouraged the reader to think about the subject, for learning formulae is not enough for any enquiring mind.

2.16. EXAMPLES

1. Two parallel but oppositely directed forces F a distance a apart constitute a couple. Show that the moment of the couple about any axis perpendicular to the plane of the forces is Fa.
2. (a) An electrically charged small body of mass m and charge $-e$ is moving in a circle of radius r round a fixed heavy particle carrying a charge $+ Ze$. Assuming that the force F between two electric charges q_1 and q_2 obeys the law $F = q_1 q_2/(4\pi\varepsilon_o r^2)$, where ε_o is the permittivity of free space, show that the speed v with which the small body is moving is given by

$$v = (Ze^2/4\pi\varepsilon_o mr)^{\frac{1}{2}}.$$

(b) The situation described in (a) above represents the picture used by Bohr in establishing a theory of atomic structure. The negatively charged particle was an electron and the heavy particle a nucleus. Bohr hypothesized though that the angular momentum l of the orbiting electron was quantized, i.e., $l = n\hbar$ where n is an integer and $\hbar = h/2\pi$ where h is Planck's constant. Show that if this quantum rule is applied, the permissible values of the electron orbit radius r_n vary as n^2 and the permissible kinetic energies of the electron vary as n^{-2}. Note that the total energy of the electron will also include potential energy; show that this potential energy is negative and has twice the magnitude of the kinetic energy.

3. A cricket ball is bowled with velocity V relative to a cricket pitch, and has imparted to it a spin of angular velocity ω about an axis at right angles to its direction of flight. Write down an expression for the kinetic energy, $\frac{1}{2}m_x v_x^2$, of each element of mass m_x of the ball, where v_x is the velocity of m_x relative to the cricket pitch. Hence show that the total kinetic energy T of the ball is given by

$$T = \tfrac{1}{2}MV^2 + \tfrac{1}{2}I\omega^2$$

where M is the mass and I the moment of inertia about the axis of spin of the ball. Note that this result holds true whatever direction the spin axis is pointing in.

4. A thin shell, in the form of a right circular cylinder, is rolling along the ground. Show that its total kinetic energy is evenly divided between translational kinetic energy and rotational kinetic energy.

5. A right circular cylinder of uniform density is rolling under gravity down a plane inclined at an angle ϕ to the horizontal. Assuming that the mass of the cylinder is M, its radius is R, and its moment of inertia about its axis is $\frac{1}{2}MR^2$, write down the equations of motion which will give its linear acceleration a and its angular acceleration α. Hence show that a is given by

$$a = \frac{2}{3} g \sin \phi,$$

where g is the acceleration due to gravity. What fraction of the cylinder's total kinetic energy is rotational kinetic energy? If the frictional force on the cylinder rim was F, show that the work done by this force after the cylinder has rolled a distance from rest of s along the plane is equal to the rotational kinetic energy. Show also that the work done by the gravitational force is equal to the total kinetic energy. What force has therefore provided the work to give the cylinder its translational kinetic energy?

6. An electron has charge of magnitude e and mass m. It is travelling in an orbit round a nucleus, which gives it an angular momentum $\hbar$ about the nucleus. ($\hbar = h/2\pi$ where h is Planck's constant). The orbit constitutes a current loop of magnetic moment $\mu = e\hbar/2m$. Show that if this loop is placed with its normal at right angles to a magnetic field B, so that it experiences a couple of moment μB, it will precess about the field direction with angular velocity Ω given by

$$\Omega = eB/2m.$$

(This is one case of the 'Larmor precession', important in atomic theory).

7. Prove the theorem of parallel axes in the following way. Consider a body of mass M whose moment of inertia about an axis through its centre of mass I_C is given by

$$I_C = \Sigma_{\text{all } x}\, m_x r_x^2,$$

where m_x is a small mass a distance r_x from the axis. Consider now another axis a distance $\underline{d}$ away, but parallel, so that the distance of m_x from $\underline{d} = \underline{d} + \underline{r}$. Write down the expression for the moment of inertia about this other axis, I_A. Remembering that

$$\Sigma_{\text{all } x}\, m_x r_x = 0$$

from the definition of centre of mass, show as required that

$$I_A = I_C + Md^2.$$

8. Prove the theorem of perpendicular axes. The proof is so simple that no hints are required.

9. Calculate the moment of inertia of a thin flat uniform square body of mass M and side a about an axis perpendicular to its plane and through its centre. This is most easily done by calculating the moments of inertia about suitable perpendicular axes in the plane of the body and then using the theorem of perpendicular axes. Deduce from the above the moment of inertia of a cube of sides a and mass M about an axis perpendicular to a face through the centre. Use the concept of ellipsoid of inertia to deduce the moment of inertia of the cube about a central axis which runs through diagonally opposing corners.

10. A man of mass 80 kg is standing at the centre of a circular platform of radius > 3 m, which is rotating at an angular velocity of one radian per second. He walks towards a point on the edge of the platform at a speed of 5 km hour^{-1}. Estimate the magnitude of (a) the centrifugal force, (b) the Coriolis force felt by him when he is (i) 20 cm, (ii) 3 m from the centre. Convert these forces from Newtons to kilograms weight to help to visualize their effect ($g = 9{\cdot}8$ m s^{-2})

3. Vibratory motion

3.1. INTRODUCTION

The last two chapters have introduced the fundamental ideas of mechanics, together with some discussion of various forms of linear and rotary motion. One form of motion not so far discussed is vibratory motion; this chapter will discuss this type of motion, which is worth particular attention as it is so widespread in nature.

Vibrations occur both on the small and on the large scale. All solids and liquids are composed of vibrating atoms; the energy of these vibrations shows itself as heat, and the hotter the solid or liquid gets, the more energetic the atomic vibrations become. More obvious are the vibrations of larger bodies than individual atoms, for example, when sound is produced by vibrating either the strings of a musical instrument or the diaphragm of a loudspeaker. Vibrations can also pose problems to the engineer, to whom they may be undesirable.

Consider now the problem of describing a vibration. Suppose a particle, initially at a point described by $x = 0$, is set into vibration in the x direction. The resulting motion could be fairly simple, as shown in the 'x,t' graph of Fig. 3.1(a), or it could be complicated, as shown in Fig. 3.1(b) and (c). All these motions have one important feature in common, namely that after a certain constant time interval, known as the period T of the vibration, they repeat themselves. The inverse of the period, i.e., the number of vibrations in unit time, is known as the frequency, ν or f. In SI units, the number of vibrations per second is the frequency in hertz (Hz).

If one has a vibration at a particular frequency and if one wishes to describe this vibration, graphs such as those of

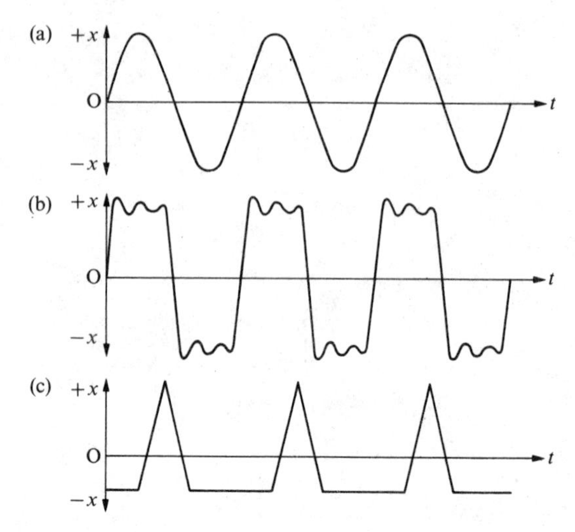

FIG. 3.1. Displacement-time graphs for various possible vibratory movements.

Fig. 3.1 may not be explicit enough for the purpose. One may wish to write down an explicit equation giving x as a function of t, and this may not be easy to do. However, a guideline as to how to set about it was given by Fourier, a French physicist and army engineer; in 1822, he put forward a theorem which, in terms suitable for use at the moment, may be stated as follows.

> 'Any regular periodic vibration, of frequency f, can be expressed as the sum of a series of simple harmonic motions whose frequencies are integral multiples of f.'

Thus an understanding of simple harmonic motion is valuable for the study of all regular vibrations; however, before this is discussed further, it is desirable to make clear what is meant by simple harmonic motion.

3.2. SIMPLE HARMONIC MOTION (SHM)

One does not get very far in the study of physics before encountering simple harmonic motion, which will from now on often be referred to as SHM. Usually, when one first encounters it, it does not seem to be particularly simple. However, when one later comes to realize the complexities which vibratory motion can display, one soon appreciates that it is the simplest type of vibration from the point of view of a mathematical description.

There are two ways of defining simple harmonic motion; they are both in common use and are equally valid. The first definition is essentially pictorial.

Definition 1. If a point is moving with uniform speed round a circle, then the motion of the foot of the perpendicular from that point on to any straight line in the plane of the circle is said to be simple harmonic.

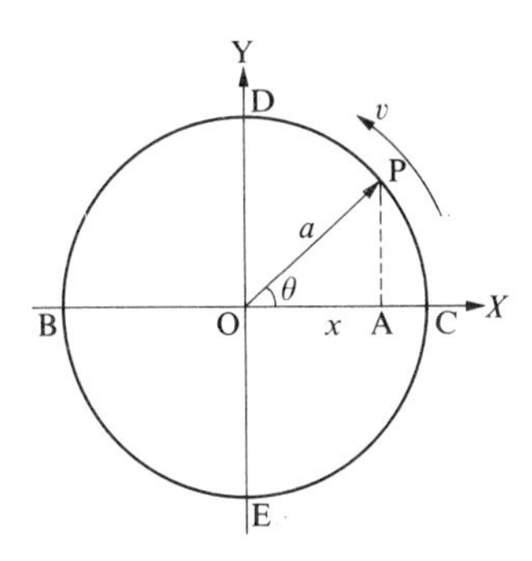

FIG. 3.2. To illustrate the first definition of simple harmonic motion. Note that θ varies with time in such a way that $\theta = \omega t + \phi$

This definition is illustrated by Fig. 3.2. Point P is moving round the circle of radius a with centre at O at the uniform speed v. The circle diameter BOC is a line in the plane of the circle. The foot of the perpendicular from point P on to that particular line is point A. As P moves round the circle, point A moves backwards and forwards along line BOC and its motion is, by definition, SHM.

To obtain a formal mathematical description of A's motion, one can describe it in terms of the distance of A from the centre of the motion O. Let this distance, AO, be x, as indicated on the figure,so that BOC becomes in effect an X axis of coordinates with origin at O; distances to the left of O are thus reckoned negative. Of course, x varies with time t, and the way it varies with t provides the description now being sought. The angle POC is varying with time; denoting it as θ, clearly

$$\theta = \omega t + \phi$$

where ϕ is the value of θ when $t = 0$, and $\omega (= v/a)$ is the angular velocity of P about O. So one can at once write (by inspection of the triangle AOP)

$$x = a \cos(\omega t + \phi). \tag{3.1}$$

Eqn (3.1) thus provides a simple formal mathematical description of a simple harmonic motion. The angle ϕ, which is only an indication of where P was when one started to measure the time t, is sometimes called the phase angle; clearly, if $\phi = 0$,

$$x = a \cos \omega t, \tag{3.2a}$$

and if $\phi = - \pi/2$,

$$x = a \sin \omega t; \tag{3.2b}$$

these two equations, (3.2a) and (3.2b), also describe SHM s. Now A is moving with a vibratory motion, whose period T and frequency f are related to ω by

$$T = 2\pi/\omega$$

and

$$2\pi f = \omega,$$

so that eqn (3.1) could equally well be written as

$$x = a \cos(2\pi ft + \phi),$$

and so on for the other equations.

It is clearly possible to produce a variety of expressions for SHM in terms of sine or cosine functions. It is desirable though that a mathematical description should always be as simple as possible, both to write down and to work with.

Experience has shown that this optimum simplicity and convenience is often achieved by treating the distance x as part of a complex number z, whose real part always has the value x. The selection of z is made easy by direct comparison of Fig. 3.2, with the line EOD regarded as a Y axis, with an Argand diagram as shown in Fig. 3.3. A suitable z is given by

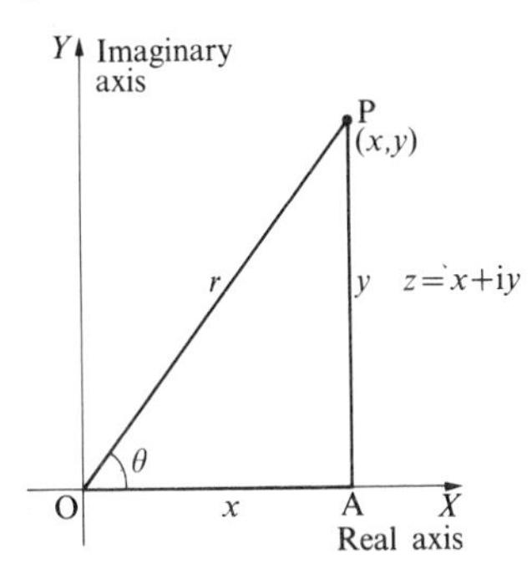

FIG. 3.3. The Argand diagram for the geometrical representation of a complex number, z, by the point P.

$$z = a \exp\{i(\omega t + \phi)\}, \tag{3.3}$$

because, as can be seen from the relation

$$\exp(i\theta) = \cos\theta + i\sin\theta, \tag{3.4}$$

the real part x of this z takes the form

$$x = a\cos(\omega t + \phi),$$

which is identical to eqn (3.1).

The advantage of using z includes simplicity when it comes to differentiation (and integration). If one wishes to find the velocity with which point A is moving, i.e. (dx/dt), one notes that, as

$$\frac{dz}{dt} = \frac{dx}{dt} + i\,\frac{dy}{dt} \tag{3.5}$$

this velocity is simply the real part of (dz/dt). Differentiating therefore eqn (3.3) with respect to time, one obtains

$$\frac{dz}{dt} = i\omega a \exp\{i(\omega t + \phi)\}, \tag{3.6}$$

which, by eqn (3.4),

$$= i\omega a\{\cos(\omega t + \phi) + i \sin(\omega t + \phi)\}$$

$$= -\omega a \sin(\omega t + \phi) + i\omega a \cos(\omega t + \phi). \tag{3.7}$$

From eqns (3.5) and (3.7), it follows at once that

$$\frac{dx}{dt} = -\omega a \sin(\omega t + \phi). \tag{3.8}$$

To obtain this result, it was not necessary to use complex numbers; a straight differentiation of eqn (3.1) gives the same result. The simplicity gained can be seen by inspection of eqns (3.6) and (3.3) which shows that

$$\frac{dz}{dt} = i\omega z.$$

Thus, in this case, the operator (d/dt) is equivalent in its action to a multiplication by $i\omega$.

If one now seeks to find the acceleration of A, d^2x/dt^2, further differentiation (or, as has been shown, multiplication by $i\omega$) of eqn (3.6) gives

$$\frac{d^2z}{dt^2} = -\omega^2 z,$$

leading to

$$\frac{d^2x}{dt^2} = -\omega^2 x, \tag{3.9}$$

an equation which also follows by differentiation of eqn (3.8).

Eqn (3.9) is important. It is the differential equation of SHM, and may be used to define SHM as follows:

Definition 2. If a particle is moving in such a way that its acceleration is always both proportional to its distance from a fixed point and also directed towards that point, then the motion of that particle is said to be simple harmonic.

Eqn (3.9) can thus be regarded as an equation which defines SHM, ω^2 being the constant of proportionality and the minus sign indicating the relative directions of acceleration (d^2x/dt^2) and distance from the fixed point, or the displacement, x.

It is instructive now to put eqn (3.9) into the form of an equation of motion of the type of eqn (1.2), i.e.

mass x acceleration = force

Multiplying both sides of eqn (3.9) by m and substituting the constant k for $m\omega^2$ gives now the equation

$$m \frac{d^2x}{dt^2} = -kx. \tag{3.10}$$

This means that if a particle of mass m is executing simple harmonic motion, then the force on the particle is proportional to the displacement of that particle from the centre of the motion and is acting towards the centre.

Conversely, it follows that if a particle experiences only such a force, i.e., one proportional to distance from a point and directed towards the point, then the particle will execute simple harmonic motion. Such forces can occur when elastic forces are involved, but Hooke's law has to be perfectly obeyed to produce a perfect SHM. This point is discussed further when anharmonic oscillations are discussed in Section 3.3.

A discussion of SHM often commences with definition 2, i.e. with an equation of the form

$$\frac{d^2x}{dt^2} = -Ax \tag{3.11}$$

where A is some constant; it then goes on to solve eqn (3.11) by conventional techniques, two constants of integration will appear in any solution obtained, and the solution can always be expressed in the form, say, of eqn (3.1), i.e.

$$x = a \cos (\omega t + \phi)$$

where $\omega = A^{\frac{1}{2}}$, and a and ϕ, amplitude and phase angle respectively, are the constants of integration; these constants can be determined by the insertion of appropriate boundary conditions, such as their values when $t = 0$. If one treats the variable in eqn (3.11) as complex, i.e. as z rather than as x, the solution could alternatively take the form of eqn (3.3).

Before leaving the topic of SHM, it should be noted that the equation

$$\frac{d^2x}{dt^2} = -Ax,$$

being equivalent to

$$\frac{d^2x}{dt^2} = -\omega^2 x,$$

leads to an oscillation whose period

$$T = \frac{2\pi}{\omega} = 2\pi(A)^{-\frac{1}{2}};$$

thus the period of the motion described by eqn (3.10) is given by

$$T = 2\pi(m/k)^{\frac{1}{2}}.$$

It is important to note that the period of SHM is entirely independent of the amplitude, a, of the motion. It is, in fact, a most important characteristic of a true SHM that the period does *not* depend on the amplitude.

The matter of which definition to use is a matter of personal choice. If definition 1 is used, one is going straight to a description of the motion being defined; some may find this preferable, and one is certainly in good company in using it (see, for example, the *Treatise on natural philosophy* by Sir

William Thomson, later Lord Kelvin, and P.G. Tait, published by the Clarendon Press, Oxford, in 1867). Definition 2 defines only the differential equation of the motion, which has to be solved in order to obtain an expression for the SHM. On the other hand, this definition draws attention to the physics of situation in which SHM can arise, emphasizing for example, the need for perfect elasticity if the restoring force is elastic in nature; this too has merit. While cases in physics can arise where choice of definition is important, this does not seem to be one of these; the definitions are equally valid, and the nature of the motion is unaffected by which one is chosen.

3.3. ANHARMONIC MOTIONS

Consider a particle of mass m undergoing the SHM described by eqn (3.10), i.e. by

$$m\frac{d^2x}{dt^2} = -kx.$$

The '$-kx$' on the right makes it clear that the force on the particle is always proportional to the distance of the particle away from the centre of the motion. Now in any real case of a vibrating particle, it is quite possible that this simple relationship between force and distance will not hold. If it does not, it will however normally be possible to express the relationship in the form of a power series so that the equation of motion of a particle would become

$$m\frac{d^2x}{dt^2} = -kx - \alpha x^2 - \beta x^3 - \gamma x^4 \ldots\ldots \quad (3.12)$$

where α, β, γ, etc. are constants. Expressing the force as a power series in the displacement x is a convenient mathematical generalization; whenever one is uncertain of the nature of the relationship between two directly related physical quantities, one can usually express one as a power series of the other and know that putting in the right value of the constants will lead to a correct solution. This is a well-known mathematical statement. It may not however always be easy to determine the constants, but once they are known the problem of describing the motion becomes that of solving eqn (3.12) using only as many terms on the right-hand side as are necessary. The series of course must be convergent for this to be generally possible.

The problem of solving eqn (3.12) is considerably more difficult mathematically than that of solving eqn (3.10). To attempt to discuss it in detail here would be to convert a discussion on mechanics into a discussion on mathematics, introducing elliptic integrals and the like. It is fully discussed in more advanced texts. The important physics is that the motion will still be regularly periodic even if it lacks the properties of SHM (e.g., the frequency will no longer be independent of the amplitude). One can nevertheless still use Fourier's theorem to to describe the motion in terms of a sum of SHM s.

Naturally occurring oscillations are very often anharmonic. In elementary discussion, the simple pendulum is often quoted as an example of SHM, with the proviso that the amplitude of the oscillation is kept small. In fact, its motion is described by eqn (3.12) rather than eqn (3.10) and is anharmonic; however, for small values of x, $-kx$ is sufficiently bigger than all the other terms in the series put together for eqn (3.12) to reduce to eqn (3.10) for practical purposes. In doing so, one is making a common-sense approximation to meet real needs for a useful solution; this is a procedure that a physicist must always be ready to adopt. For most practical purposes, the simple pendulum at low amplitude moves with SHM.

Sometimes however the anharmonicity of an oscillation can have important practical consequences. The anharmonicity of the oscillations of the atoms in a solid affects significantly the thermal properties of the solid, such as the thermal expansion. In the case of thermal explansion, the law of force between atoms is such that, when the amplitude of atomic oscillations increases, the atoms tend to move further apart. For any one atom, this means a movement of its centre of oscillation. Change of amplitude will produce a movement of the centre of oscillation whenever the coefficients of the even powers of x in eqn (3.12) are other than zero, and, in the case of solids, this movement leads to thermal expansion.

Whenever the force and the displacement, or equivalent quantities, are not directly proportional to one another, i.e., whenever the system is non-linear, the effects of anharmonicity can become evident and can be of sufficient importance to be put to practical use. Non-linear phenomena are of current research interest; as with the simple pendulum, they often become important only when the amplitude of the motion being studied becomes relatively large, but they can also be important at quite low amplitudes. One important effect is that when two or more oscillations of different frequencies mix together in a non-linear medium, sum and difference frequencies can be created and this effect can have useful applications as well as being of theoretical significance. One concludes that, taken overall, anharmonic effects are by no means always trivial.

3.4. COMPLICATED VIBRATIONS AND FOURIER ANALYSIS

In Section 3.1, Fourier's theorem was stated in the following terms:

'Any regular periodic vibration of frequency f can be expressed as the sum of a series of simple harmonic motions whose frequencies are the integral multiples of f.'

One way of expressing a SHM is that given as eqn (3.1),

namely

$$x = a \cos (\omega t + \phi).$$

Using the standard expression for the cosine of the sum of two angles, i.e. $\cos(\alpha + \beta) = \cos \alpha \cos \beta - \sin \alpha \sin \beta$, and then substituting A for $a \cos \phi$ and B for $-a \sin \phi$, it follows that

$$x = A \cos \omega t + B \sin \omega t. \tag{3.13}$$

This equation, like eqn (3.1), can always be obtained by solving eqn (3.11), A and B now being the two constants of integration each containing the amplitude and phase.

Let X be the displacement in a complicated vibration of angular frequency ω ($= f/2\pi$). Then, by Fourier's theorem, X is equal to the sum of a number of SHM s of the general form, using eqn (3.13),

$$x_n = A_n \cos n\omega t + B_n \sin n\omega t, \tag{3.14}$$

where n takes the values 0, 1, 2 ----- etc. This sum may be written in the following way

$$X = A_o + \sum_{n=1}^{\infty} A_n \cos n\omega t + \sum_{n=1}^{\infty} B_n \sin n\omega t. \tag{3.15}$$

The right-hand side of eqn (3.15) is the Fourier series describing X. The term A_o indicates a tendency for the displacement to be more in one direction than the other, and for many vibrations it is zero. One might not appreciate this until one had carried out a Fourier analysis, i.e., evaluated the terms on the right-hand side of eqn (3.15).

The nth term in the series, as expressed by eqn (3.14), is known as the nth harmonic. One can see at once why a vibration which contains only the first harmonic is described as simple.

Complicated vibrations occur in many branches of physics. In sound, for example, the note produced by a musical instrument can be quite complicated; for this particular case, the first harmonic is often called the fundamental and the nth harmonic is called the $(n - 1)$th overtone. Fourier in fact developed his theory in connection with heat flow, being interested in how heat flowed away from a source whose temperature varied periodically, such as the flow of heat into the earth from its surface, periodically warmed by sunlight.

3.5. ENERGY IN A SIMPLE HARMONIC MOTION

Consider a particle of mass m oscillating along a line with SHM of amplitude a and angular frequency ω. Let its displacement from the centre be x and the motion be described by eqn (3.1), i.e.

$$x = a \cos (\omega t + \phi),$$

which is the motion of point A on Fig. 3.2. At any stage in the oscillation, the kinetic energy T of the particle is clearly given by

$$T = \tfrac{1}{2}\, m \left(\frac{dx}{dt}\right)^2,$$

which, from eqn (3.9) means that

$$T = \tfrac{1}{2}\, m\omega^2 a^2 \sin^2 (\omega t + \phi). \tag{3.16}$$

One notes that at the ends of the oscillation, $\sin(\omega t + \phi) = 0$, so $T = 0$, and at the centre of the oscillation, $\sin(\omega t + \phi) = 1$, so that

$$T = \tfrac{1}{2} m\omega^2 a^2. \tag{3.17}$$

Eqn (3.17) is obvious when one notes that $v = \omega a$ and the speed of the particle is v at the centre (see Fig. 3.2). There will also be potential energy associated with the oscillation; this will be the work done against the restoring force to get the particle from the centre to any given place. Referring to eqn (3.10), one notes that the restoring force is kx, where $k = m\omega^2$, so the work done

$$= \int_0^x m\omega^2 x\,dx = \tfrac{1}{2}\, m\omega^2 x^2 .$$

Equating this to the potential energy V of the particle, while substituting for x from eqn (3.1), it follows that

$$V = \tfrac{1}{2} m\omega^2 a^2 \cos^2(\omega t + \phi). \tag{3.18}$$

Here again, it is clear that $V = 0$ at the centre of the oscillation, and at the ends

$$V = \tfrac{1}{2} m\omega^2 a^2 . \tag{3.19}$$

The complete conversion of kinetic energy into potential energy, as shown by eqns (3.17) and (3.19), are what one would expect; in general, the total energy E will be $T + V$, which from eqns (3.16) and (3.18) means

$$E = \tfrac{1}{2}\, m\omega^2 a^2 \{\sin^2(\omega t + \phi) + \cos^2 (\omega t + \phi)\};$$

the term in braces clearly equals one, so it follows that the energy E in this simple harmonic motion is given at all times by

$$E = \tfrac{1}{2} m\omega^2 a^2 . \tag{3.20}$$

The constancy of the total energy in the motion should be noted; this is a feature of motion in a region where the potential energy is a clearly defined single-valued function of position, and the only force acting comes from the potential energy gradient; SHM is not the only type of motion to which this constancy applies.

3.6. DAMPED VIBRATIONS

In practice, a real mechanical system, such as a simple pendulum, set into oscillation and left to itself, experiences damping; the amplitude of the oscillations decreases and the oscillations appear slowly to die out. Frictional effects of some sort are slowing the oscillation down.

Suppose then one has a particle of mass m which is oscillating; if the oscillations are SHM, the equation of motion is eqn (3.10), i.e.,

$$m\frac{d^2x}{dt^2} = -kx.$$

Oscillations described by this equation retain a constant amplitude; the energy remains constant. If however the amplitude decreases, energy is being dissipated, and this energy loss must come from the work done against some frictional force. Let this force be F, some function of x and t. This force F will always act as a decelerating force; one must therefore replace eqn (3.10) by an equation of motion of the form

$$m\frac{d^2x}{dt^2} = -kx - F(x,t). \tag{3.21}$$

To calculate the form of the damped motion, one has to solve eqn (3.21). The method of solving will depend on the nature of F, i.e., what type of function of x and t it is. Sliding friction forces are usually constant, and if the friction were of this type one would treat F as a constant in eqn (3.21). In the case of a simple pendulum in air, and in a number of other possible mechanical vibrating systems, the damping force arises from the resistance of the air or other fluid to the motion. Such a force often arises from the viscosity of the fluid, when it is then termed a viscous force. It is characteristic of a viscous force that it is proportional only to the velocity with which the obstructed body is moving, provided that the temperature remains constant. Under these conditions, one can write for F,

$$F(x,t) = R\frac{dx}{dt}$$

where R is a constant. Incorporating this into eqn (3.21) and rearranging, the equation of motion becomes

$$m \frac{d^2x}{dt^2} + R \frac{dx}{dt} + kx = 0. \tag{3.22}$$

This is a straightforward second-order linear differential equation with constant coefficients, and there exist a number of alternative ways of solving it, all of which will lead to the same solution. One method is to assume that a solution exists of the form

$$x = C \exp(\lambda t). \tag{3.23}$$

If one assumes this, it follows that

$$\frac{dx}{dt} = C\lambda \exp(\lambda t) = \lambda x, \tag{3.24}$$

and

$$\frac{d^2x}{dt^2} = C\lambda^2 \exp(\lambda t) = \lambda^2 x. \tag{3.25}$$

Incorporating eqns (3.24) and (3.25) into eqn (3.22) and dividing by x, it follows that

$$m\lambda^2 + R\lambda + k = 0. \tag{3.26}$$

Eqn (3.26) is a standard quadratic equation for λ, for which the well-known solution is

$$\lambda = -\{R \pm (R^2 - 4km)^{\frac{1}{2}}\}/2m. \tag{3.27}$$

If the damping is small, $R^2 < 4km$; in this case, one can write, bringing the $2m$ within the brackets.

$$\lambda = -R/2m \pm i(k/m - R^2/4m^2)^{\frac{1}{2}}. \tag{3.28}$$

If one now recognizes that there are two possible values of λ, depending on the choice of the plus or minus sign, and that the solution can be a mixture of the two possibilities, one can incorporate eqn (3.28) in eqn (3.23) to obtain

$$x = \exp\,(-Rt/2m)\left[A\,\exp\,\{\mathrm{i}(k/m - R^2/4m^2)^{\frac{1}{2}}t\} + B\,\exp\,\{-\mathrm{i}(k/m - R^2/4m^2)^{\frac{1}{2}}t\}\right], \qquad (3.29)$$

where the constant C has been absorbed in the two constants A and B. One now has a complex solution; clearly for x, one should have written z, for x is only the real part. Using eqn (3.4) one can separate real and imaginary parts in the exponential terms, and at the same time allow for the constants A and B to be complex. The upshot of the necessary algebra is that, taking real parts only, eqn (3.29) reduces to the form

$$x = C\,\exp(-Rt/2m)\,\cos\,\{(k/m - R^2/4m^2)t + \phi\}. \qquad)3.30)$$

The two constants C and ϕ can be determined from the values of x and $(\mathrm{d}x/\mathrm{d}t)$ when $x = 0$.

Eqn (3.30) is recognizable as similar to eqn (3.1). The solution is an oscillation, and the differences from SHM are

(a) the amplitude is no longer the constant a, but has a value $C\,\exp\,(-Rt/2m)$ which represents an exponential decay with time, and

(b) the angular frequency ω is now given by the equation

$$\omega = (k/m - R^2/4m^2)^{\frac{1}{2}}.$$

For the undamped motion, $\omega = (k/m)^{\frac{1}{2}}$, so the frequency is slowed down by the damping; the values are such that this change of frequency is a smaller and less noticeable effect than the amplitude decrease in most practical cases. One notes that, if $R = 0$, i.e., if the damping is removed, eqn (3.30) becomes at once identical to eqn (3.1)

If the damping is greater, then $R^2 \geqslant 4km$; from eqn (3.27), λ is always real. Oscillations can no longer arise, and a displaced system with zero initial velocity will return without any trace of oscillation to its undisplaced state, the approach becoming exponential in the end. Fig. 3.4 illustrates the possibilities.

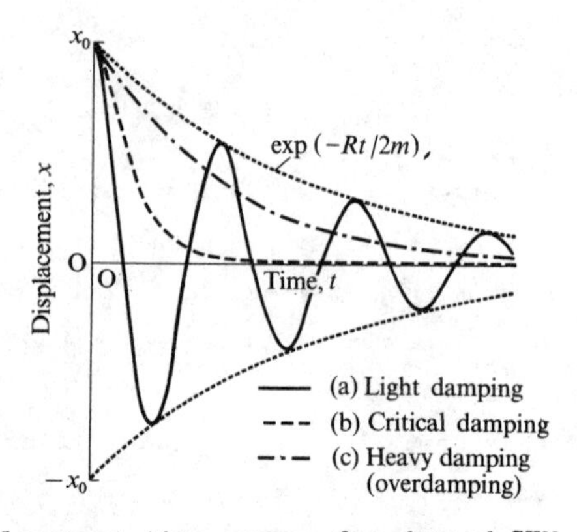

FIG. 3.4. Displacement-time curves for damped SHM. The initial conditions in each case are that, when $t = 0$, $x = x_0$ and $dx/dt = 0$

The condition $R^2 = 4km$, when the approach is as fast as it can be without any overshoot or sign of oscillation, is sometimes called critical damping.

To sum up this section, it can be concluded in real cases, damping is often of importance, and the exact behaviour will depend on the form of the damping force and its dependence on velocity, etc. The case of a viscous damping force is not uncommon and its solution has been discussed. The equations occur in other branches of physics, notably in electrical circuit theory, with similar solutions.

3.7. FORCED VIBRATIONS AND RESONANCE

A phenomenon, which can be of considerable importance to the engineer in the case of large-scale mechanical vibrations, arises from the forcing of the vibration and is known as resonance. By forcing the vibration, one means adding a periodic force; every child who has played on a swing knows that he or she can force the vibration and so build up its amplitude. For convenience in solving the mathematics, let it be supposed that a periodic force consisting of the real part of $F \exp(i\omega t)$ be added to the forces on the right-hand side of eqn (3.21) so that, after rearranging, one obtains

$$m\frac{d^2x}{dt^2} + R\frac{dx}{dt} + kx = F\exp(i\omega t). \qquad (3.31)$$

This is now the equation of motion for a particle of mass m moving with a damped SHM driven by an oscillating force. The angular frequency of the force, ω, is not necessarily the same as the natural undamped frequency of the SHM, $\omega_0 = (k/m)^{\frac{1}{2}}$. To solve this, one can again guess at a solution. A damped SHM dies off with time, so once the SHM has died off, one might expect m to vibrate with the same period as the driving force. Therefore one possible solution might take the form

$$x = A\exp(i\omega t) \qquad (3.32)$$

where A may be complex but where one understands that it is only the real part of the right-hand side which represents x. Differentiating eqn (3.32) with respect to time, one obtains

$$\frac{dx}{dt} = i\omega x; \quad \frac{d^2x}{dt^2} = -\omega^2 x.$$

Substituting for x, etc., from the above into eqn (3.31) and cancelling $\exp(i\omega t)$ from both sides of the result, one obtains

$$(-m\omega^2 + i\omega R + k)A = F.$$

It follows therefore that

$$A = \frac{F}{(k - m\omega^2) + i\omega R}$$

To separate the real and imaginary parts of A, one multiplies top and bottom by $\{(k - m\omega^2) - i\omega R\}$ to obtain

$$A = \frac{F(k - m\omega^2)}{(k - m\omega^2)^2 + \omega^2 R^2} - \frac{iF\omega R}{(k - m\omega^2)^2 + \omega^2 R^2}. \qquad (3.33)$$

It is now desirable to convert everything back to real terms. The real part of the driving force $F \exp(i\omega t)$

$$= F \cos \omega t. \tag{3.34}$$

In finding the real part of the right-hand side of eqn (3.32), one notes that

$$\exp(i\omega t) = \cos \omega t + i \sin \omega t. \tag{3.35}$$

Multiplying then the right-hand sides of eqns (3.33) and (3.35) and taking only the real parts, it follows that

$$x = \frac{F(k - m\omega^2)}{(k - m\omega^2)^2 + \omega^2 R^2} \cos \omega t + \frac{F\omega R}{(k - m\omega^2)^2 + \omega^2 R^2} \sin \omega t. \tag{3.36}$$

The solution of eqn (3.31) given by eqn (3.36) represents a particular integral; to obtain the full solution, one must add the complementary function which is the solution of eqn (3.22) and which has been discussed therefore in Section 3.6. The part of the full solution represented by the complementary function dies out with time. However, if a damped system is suddenly set into forced vibration, this complementary function part may show itself in the motion; it is then referred to as a transient. Such transients can be important (e.g., in the production of musical notes), but the steady-state solution, eqn (3.36), is the one to which the system will settle down.

One can express the steady-state solution in the form

$$x = A \cos(\omega t + \phi) \tag{3.37}$$

To evaluate A and ϕ in eqn (3.37), first expand it to give

$$x = A(\cos \omega t \cos \phi - \sin \omega t \sin \phi) \tag{3.38}$$

By now equating the coefficients of $\cos \omega t$ and $\sin \omega t$ in eqns (3.38) and (3.36), one finds that

$$\tan \phi = \omega R/(m\omega^2 - k), \tag{3.39}$$

and

$$A = F/\{(k - m\omega^2)^2 + \omega^2 R^2\}^{\frac{1}{2}}. \qquad (3.40)$$

These two results can be incorporated into eqn (3.37). It is worthwhile looking a little into what they mean in terms of how the motion of the mass m varies with the frequency ω of the driving force relative to the natural undamped frequency, $\omega_o = (k/m)^{\frac{1}{2}}$. One notes that $(m\omega^2 - k)$ can be rewritten as $m(\omega^2 - \omega_o^2)$, and that this term will tend to zero as ω approaches ω_o. For ω very small, $\phi \to o$, i.e., the displacement and applied force are in phase. As ω increases, ϕ becomes negative indicating that the displacement is lagging in phase behind the driving force; when $\omega = \omega_o$, eqn (3.39) makes it clear that this lag is then $\pi/2$ in phase, i.e., the applied force is reaching a maximum every time the displacement goes through zero, and the phase lag means that it is encouraging acceleration in the vibration in such a way as to increase the amplitude. When $\omega = \omega_o$, it turns out that the velocity of m reaches maximum values and the condition is known as velocity resonance. The amplitude is near its maximum but not quite there. (In the electrical analogue, the resonant circuit, it is the equivalent to velocity resonance which is called resonance). As ω increases above ω_o, the lag of displacement increases reaching π when $\omega >> \omega_o$. Under these conditions, the displacement and forcing oscillations are exactly out of phase and the amplitude has decreased to a relatively small value.

Energy is most efficiently transferred from the driving force to the oscillating mass under resonance conditions; this situation can be important in many physical phenomena. Fig. 3.5 illustrates some of these characteristics of resonance.

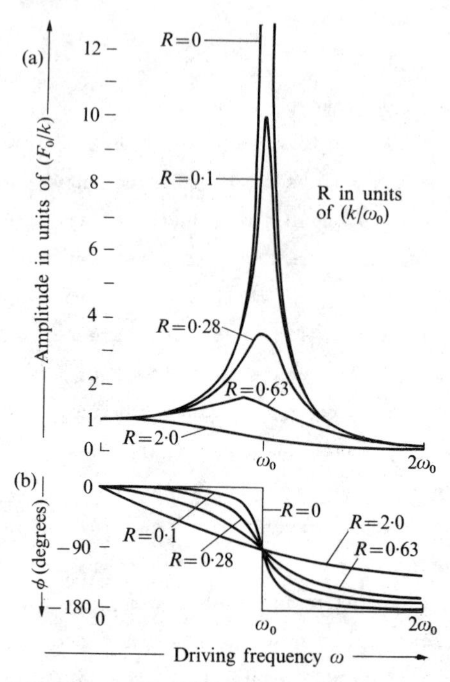

FIG. 3.5. To illustrate amplitude resonance. (a)Amplitude as a function of driving frequency for various R, eqn (3.40), and (b) corresponding phase angle, eqn (3.39). Note that no amplitude resonance is seen if $R > 2^{\frac{1}{2}}(k/\omega)$.

3.8. CONCLUSIONS

This section has been a very brief review of the formal mathematical treatment of vibrations. The applications of this theory are widespread. The phenomenon of thermal expansion has already been mentioned as one to which vibration theory may be applied in the search for an explanation; however, vibration theory has broader application than this in solid-state physics, for the theoretical explanation of any thermal property (such as, for example, thermal conduction) is likely to need vibration theory. Vibrations of atoms occur even near the absolute zero of temperature; helium, for example, remains a liquid right down to absolute zero because its atoms vibrate too much for it to settle down as a solid! It takes more than the classical theory of vibrations to account for this; one has to invoke quantum theory to obtain a solution.

However, no quantum theory is needed for large scale phenomena. The seismologist who is seeking to detect earth vibrations caused by some distant earthquake needs only classical theory in designing his apparatus and in interpreting his results. The acoustics consultant, who is called in to minimize the vibrations caused by some large unruly machinery, needs to understand the classical theories of vibration and damping to achieve his object.

Nor does there need to be some externally obvious physical vibration for the theory to be necessary and relevant; the electronics engineer, when dealing with tuned and resonant circuits, uses the identical mathematical theory to discuss and interpret the behaviour of electric currents in these circuits.

The understanding of vibration theory thus has widespread applications in science and engineering, and its importance is not to be ignored.

3.9. EXAMPLES

1. Let the time between dawn and dusk be L. It is then found that, in London, L varies with time t during the year very approximately according to the equation

$$L = 12 - 4 \cos \omega t \text{ hours},$$

where $t = 0$ around December 22 and the period of the SHM term is approximately 365 days. By how much is L changing (a) in the 24 hours immediately following the time when $\omega t = 0$ or π, (b) in a 24 hour period when the daylight hours are changing in length at their fastest rate?

2. A body is suspended from a thin straight elastic wire

fixed to a rigid support. The line of the wire is vertical and, if the body is twisted from its equilibrium position through an angle θ about the wire as axis, the wire will exert on the body a restoring torque $k\theta$. The moment of inertia of the body about the axis of twist is I. If, after twisting, the body is released, show that it will execute torsional oscillations which will be SHM of period $T = 2\pi(I/k)^{\frac{1}{2}}$.

3. A mass M is suspended at the lower end of a vertical spring, whose upper end is fixed to a rigid support. Before the mass was attached, the spring had a length l; after the attachment, it extended to $l + x$. Assuming that the spring obeys Hooke's law, show that, if the mass is pulled down a short extra distance and released, it will oscillate up and down with SHM of period $T = 2\pi(x/g)^{\frac{1}{2}}$.

4. In order to accurately measure g, the acceleration due to gravity, it is sometimes necessary to use a rigid pendulum. This can take the form of a long bar, whose moment of inertia about an axis through its centre of mass and perpendicular to its length can be written as Mk^2. The bar is set up so that it can oscillate as a pendulum about an axis parallel to this axis a distance d along the bar. Show that, for small oscillations, the period of oscillation T is given by

$$T = 2\pi\left(\frac{k^2 + d^2}{2dg}\right)^{\frac{1}{2}}.$$

Hence show that, for any one value of T, there may be two possible values of d, and sketch the form of the T-d graph. For any oscillation of period T, the length l of the simple equivalent pendulum is obtained from the equation $T = 2\pi(l/g)^{\frac{1}{2}}$. If d_1 and d_2 are two distances from the centre of mass of axes of oscillations which have the same period T, show that $l = d_1 + d_2$.

5. A complicated vibration has a displacement A which is a function of time $A(t)$. This function is symmetrical about $t = 0$, i.e., $A(t) = A(-t)$. $A(t)$ is now expanded as a Fourier series. Explain why there are no 'sine' terms in this series.

6. A body of mass m is undergoing anharmonic vibrations in the x direction, for which the equation of motion takes the form

$$m\frac{d^2x}{dt^2} = -kx - \alpha x^2 - \beta x^3 - \gamma x^4 - \delta x^5 - \varepsilon x^6 \text{-----}.$$

Explain why it is that, if the vibrations are symmetrical about $x = 0$, the coefficients of all the terms containing even powers of x (i.e., $\alpha, \gamma, \varepsilon$ etc) are zero.

7. A mass m is being whirled on the end of a string in a horizontal circle of radius r with uniform angular velocity ω. Show that the energy of the mass is identical to the energy which it would have were it oscillating with SHM of amplitude r and angular frequency ω.

8. The displacement x in a lightly damped SHM is given by

$$x = C \exp(-Rt/2m) \cos\{(k/m)^{\frac{1}{2}}t + \phi\}.$$

This is eqn (3.30) with $(R^2/4m^2)$ assumed to be $<<$ (k/m) as the damping is light . The maximum amplitudes of x taken on one side of the oscillation in two consecutive oscillations are A_r and A_{r+1}. Show that the logarithmic decrement δ of the oscillation, defined by

$$\delta = \ln(A_r/A_{r+1}),$$

is given by

$$\delta = 2\pi R(mk)^{-\frac{1}{2}}.$$

9. A particle of mass m obeys the equation of motion of damped driven SHM, eqn (3.31). The steady-state solution for the displacement x of the particle from its centre of oscillation is from eqns (3.37) and (3.40)

$$x = \left[F/\{(k - m\omega_o^2)^2 + \omega_o^2 R\}^{\frac{1}{2}}\right] \cos(\omega_o t + \phi).$$

Show that the condition for maximum amplitude is

$$\omega_o^2 = \omega^2 - (R^2/2m^2),$$

and that the condition for maximum velocity (i.e., velocity resonance) is

$$\omega_o^2 = \omega^2.$$

Is the energy in the motion a maximum at either amplitude resonance or velocity resonance? If so, at which?

4. Wave motion

4.1. INTRODUCTION

To the layman, the expression 'wave motion' is likely to be associated with waves on the surface of water, such as waves on the sea, or ripples on a pond. In fact, ripples on a pond provide a very good starting point for a discussion of travelling waves because they are so familiar. A stone thrown into a calm pond sets up a vibration on the water surface at its point of impact. Circular ripples, centred on this point, then spread out over the water surface. The movement of these ripples constitutes a good example of a travelling wave, which involves the transfer of vibration from one region to another. After the motion has passed, any small object floating on the water surface will be seen to have remained where it was before the wave passed. So the vibration moves through the medium carrying it without taking the medium along with it.

Another commonly encountered form of wave motion is the sound wave. The voice of a speaker sets up a vibration in the air; this vibration travels through the air away from the speaker at a speed of about 330 $m\,s^{-1}$ and is heard by a listener as a consequence of its action on his ear.

Just as the study of vibrations forms an essential part of physics, so too does the study of the transmission of these vibrations, i.e., the study of wave motion.

4.2. EXPRESSIONS FOR A TRAVELLING WAVE

Consider Fig. 4.1. At point A a vibration or disturbance describable by the expression $y_0 = f(t)$ is taking place. This vibration is being transmitted as a wave motion from A

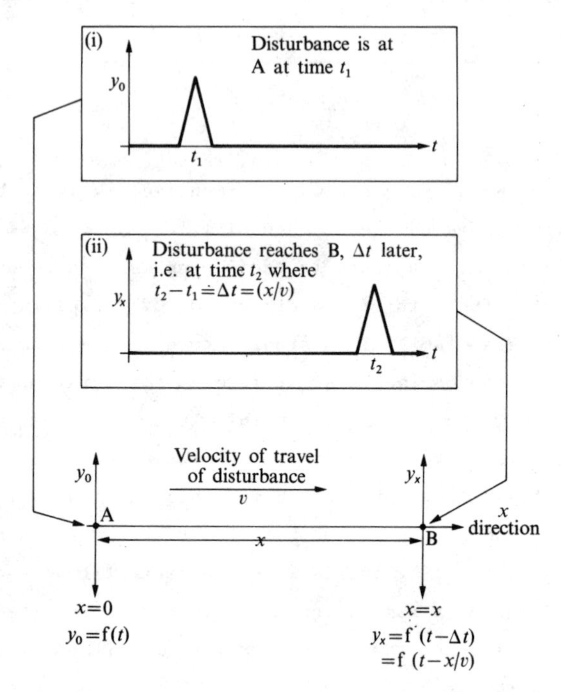

FIG. 4.1. To show the origin of the expression for a travelling wave.

(at $x = 0$) to point B (at $x = x$) with a velocity v. It takes a time (x/v) to get there. Thus the vibration that takes place at B is the vibration that took place at A a time (x/v) earlier. Such a vibration would be describable by the expression

$$y_x = f\left(t - \frac{x}{v}\right). \tag{4.1}$$

This expression is the most general form in which one can describe a vibration travelling unaltered in the direction of positive x with velocity v; it gives the disturbance or displacement y at any point along x in terms of x and t. It is important to note that it is only applicable provided that the form of the function 'f' is not changing as the wave progresses. Media through which waves travel without altering the form of 'f' are, as will shortly be explained, known as non-dispersive media; many media are dispersive and the form of 'f' can change with distance.

If the expression $y_o = f(t)$ is restricted to cover only a simple harmonic motion, so that the motion at point A on Fig. 4.1 can take the form of, say, eqn (3.2b), i.e.,

$$y_o = a \sin \omega t,$$

then the wave motion is that of a simple harmonic wave, describable by the equation

$$y_x = a \sin \omega\left(t - \frac{x}{v}\right). \qquad (4.2)$$

Alternatively, the frequency f can be introduced and eqn (4.2) becomes

$$y_x = a \sin 2\pi f\left(t - \frac{x}{v}\right). \qquad (4.3)$$

A curve of y_x plotted against x could then take the form shown in Fig. 4.2. This shows that the regular periodicity of y in

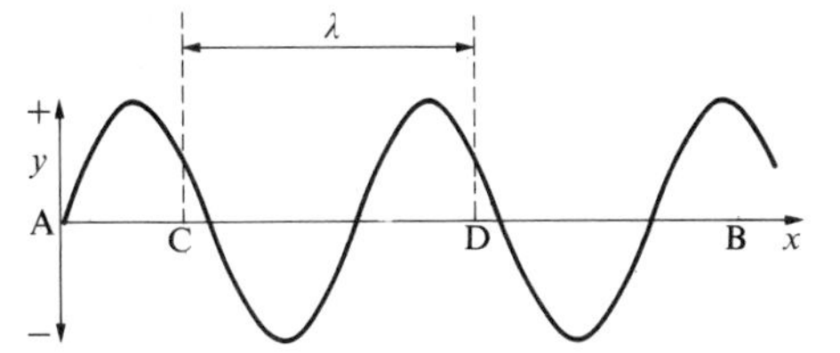

FIG. 4.2. A simple harmonic wave of wavelength λ.

time has also become a regular periodicity in space. The distance CD on the figure shows the regular periodic length, and this is customarily known as the wavelength and denoted by the symbol λ. As the wave moves past a point, such as point B, in every unit of time f vibrations take place and the wave advances a distance f times λ. The distance advanced in unit time being the velocity, it follows that

$$v = f\lambda, \tag{4.4}$$

an equation committed to memory by every aspiring physicist.

Eqns (4.2) and (4.3) are not, of course, the only forms in which the simple harmonic wave can be written. The use of eqn (4.4) and the substitution of the period T for $1/f$ leads to the form

$$y_x = a \sin 2\pi \left(\frac{t}{T} - \frac{x}{\lambda}\right). \tag{4.5}$$

In addition to wavelength, the wave vector k or wave number is frequently used. While certain physicists (particularly spectroscopists) define this quantity as the reciprocal of wave length, the more common and more generally useful definition takes the form

$$k = 2\pi/\lambda,$$

and this definition will be used here. Using it, eqns (4.2), (4.3), and (4.4) can be rewritten as

$$y_x = a \sin (\omega t - kx). \tag{4.6}$$

It is worth noting, in passing, that eqn (4.4) can now be rewritten in the form

$$v = \omega/k, \tag{4.7}$$

an equation also worth remembering.

As with simple harmonic motion, cosine can be substituted for sine, etc., and it is often most useful to use the exponential form and write

$$y_x = a \exp i (\omega t - kx). \tag{4.8}$$

This form of the equation is very commonly used when discussing the movement of sound vibrations and of electrons through solids.

For the next few sections, waves will be discussed solely as simple harmonic waves, to avoid the complications of change of waveform with distance brought about by dispersion, itself to be discussed in Section 4.11.

4.3. LONGITUDINAL AND TRANSVERSE WAVES

Fig. 4.2 is an example of what is known as a transverse wave; the vibration is taking place at right angles to the direction of motion of the wave. A wave motion produced by shaking one end of a long stretched rope from side to side would be a transverse wave. It turns out also that electromagnetic waves such as light or radio waves are transverse; the vibrating quantities in these cases are the electric and magnetic fields.

Very often however, the vibration in a wave motion is taking place in the same direction as the wave is travelling; referring to Fig. 4.2, this means that the directions of y and x are the same. Such a wave is termed longitudinal. Sound waves in a fluid are an example of such a longitudinal wave.

Most wave motions are either longitudinal or transverse. Occasionally the vibration is more complex; for example, the water waves which one sees at the seaside are associated with a vibration of the water surface that is a combination of longitudinal and transverse motions.

4.4. INTENSITY OF A WAVE

A vibration has energy; when the vibration moves, it carries its energy with it. Thus a wave motion carries energy in the direction in which it is travelling. Imagine a unit area placed normal to this direction of travel. Then the energy carried by the wave through this area in unit time is called the intensity of the wave; this intensity is measured in watts m^{-2}.

4.5. THE WAVE EQUATION

When discussing SHM, it was shown that an equation of the

form of eqn (3.11), i.e.,

$$\frac{d^2y}{dt^2} = -Ay$$

would represent SHM, so that its appearance allowed one to see at once that y executed SHM. A similar equation for a wave motion will now be derived; its appearance similarly tells one that a wave motion is present. The equations so far given (4.2), (4.3), (4.5), (4.6), and (4.8) are solutions to this equation; starting then with one of them, namely eqn (4.2), one writes, dropping the inferior x for convenience

$$y = a \sin \omega\left(t - \frac{x}{v}\right). \tag{4.9}$$

Differentiating with respect to t, keeping x constant, it follows that

$$\left(\frac{\partial y}{\partial t}\right)_x = a\omega \cos \omega\left(t - \frac{x}{v}\right)$$

and

$$\left(\frac{\partial^2 y}{\partial t^2}\right)_x = -a\omega \sin \omega\left(t - \frac{x}{v}\right)$$

$$= -\omega^2 y. \tag{4.10}$$

This result is rather what one expects; it simply indicates SHM at any fixed value of x. Now differentiating eqn (4.9) with respect to x, keeping t constant, one obtains

$$\left(\frac{\partial y}{\partial x}\right)_t = -\frac{a\omega}{v} \cos \omega\left(t - \frac{x}{v}\right)$$

and

$$\left(\frac{\partial^2 y}{\partial x^2}\right)_t = -\frac{a\omega^2}{v^2} \sin \omega\left(t - \frac{x}{v}\right)$$

$$= -\frac{\omega^2}{v^2}\, y\ . \tag{4.11}$$

This equation is of course analogous to eqn (4.10), with the wave vector $k(=\omega/v)$ replacing ω, and x replacing t.
From equations (4.10) and (4.11), it follows that

$$\left(\frac{\partial^2 y}{\partial t^2}\right)_x = v^2\left(\frac{\partial^2 y}{\partial x^2}\right)_t, \tag{4.12}$$

and this is the wave equation which one set out to derive.

If this equation appears, it is at once clear that not only does y propagate as a wave, but also that the velocity of the wave is incorporated into the equation. The derivation of the equation for a particular case is often the method of calculating the velocity of the wave in that case; this point is illustrated by the example of the next section.

4.6. VELOCITY OF TRANSVERSE WAVES ALONG A WIRE OR STRING

One form of mechanical motion that is readily seen to be a wave motion is a transverse vibration moving along a wire or rope; if a length of clothes rope, say, is fixed at one end and held out horizontally, and the unfixed end is given an up-and-down or side-to-side vibration, this vibration can be seen to travel along the rope towards the fixed end. The calculation of the velocity with which this vibration travels is relatively simple.

Fig. 4.3 shows a small section of the rope, of length δx, which, before the vibration arrived, was at the position AB. At some stage in the vibration, let it be supposed that it is displaced to the position A'B'. Let the rope be under a constant tension T and let it have mass σ per unit length. Let the transverse displacement at point x, i.e., AA', be y, and let the angles θ and ϕ be as indicated in the diagram.

Consider now the motion only in the transverse direction. The downward force on the section is the downward component of

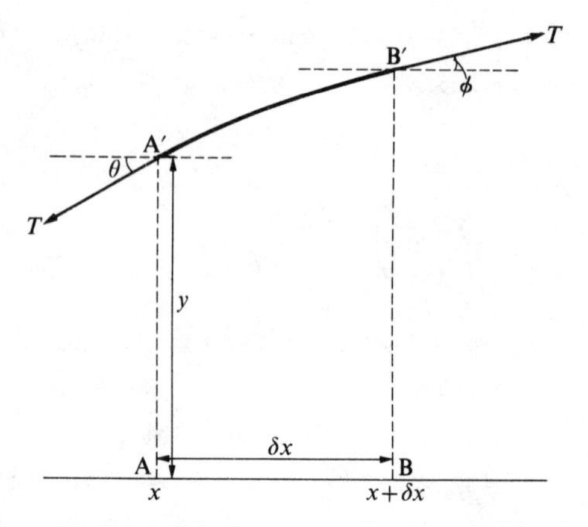

FIG. 4.3. A small section of rope, A'B', displaced during the passage of a transverse wave motion.

T acting at A', i.e., it equals

$$T \sin \theta .$$

Provided that the displacements are small, $\sin \theta \rightarrow \theta \rightarrow \left(\frac{\partial y}{\partial x}\right)_t$.

The downward force is then

$$T\left(\frac{\partial y}{\partial x}\right)_t . \qquad (4.13)$$

Similarly, the upward force is

$$T \sin \phi ,$$

i.e.,

$$= T\left\{\left(\frac{\partial y}{\partial x}\right)_t + \frac{\partial}{\partial x}\left(\frac{\partial y}{\partial x}\right)_t \delta x\right\} \qquad (4.14)$$

where one is taking into account the curvature of the section during the displacement by allowing for the change in $(\partial y/\partial x)_t$ in moving through the distance δx. The total force on the element in the transverse direction follows by subtracting eqn (4.13) from eqn (4.14), and is clearly

$$= T\left(\frac{\partial^2 y}{\partial x^2}\right)_t \delta x$$

in an upward direction.

Now the mass of the section is $\sigma\delta x$, and the acceleration is $(\partial^2 y/\partial t^2)_x$, so the equation of motion, i.e. eqn (1.2), takes the form

$$T\left(\frac{\partial^2 y}{\partial x^2}\right)_t \delta x = \sigma\delta x\left(\frac{\partial^2 y}{\partial t^2}\right)_x. \tag{4.15}$$

The similarity of eqn (4.15) to eqn (4.12), the wave equation, is already obvious; rearranging it becomes

$$\left(\frac{\partial^2 y}{\partial t^2}\right)_x = \frac{T}{\sigma}\left(\frac{\partial^2 y}{\partial x^2}\right)_t,$$

whence it follows that the velocity v of the transverse wave down the rope is given by the equation

$$v = \left(\frac{T}{\sigma}\right)^{\frac{1}{2}}. \tag{4.16}$$

This last equation, eqn (4.16), has considerable relevance in the theory of the vibrations of wires or strings in musical instruments such as the piano, violin, harp, or guitar. This point will be referred to again later when standing waves are discussed.

4.7. THE WAVE FRONT

When a wave motion extends into two or three dimensions, as do the ripples on the surface of a pond or the sound waves from any source of noise, then a line or sheet of constant phase in the wave motion is termed a wave front. A stone thrown into a pond produces circular wave fronts on the water surface; a point source of sound would produce spherical wave fronts. However, there are ways of producing other forms of wave front; one such form is the plane wave front; the expression plane wave is often used to describe such a wave with such a wave front;

a parallel beam often means a region containing plane waves. In a plane wave, one only needs to consider motion of the wave along one direction. A plane wave is thus convenient for theoretical discussion as it keeps the algebra to a minimum; in nature, it is often difficult, if not impossible, to produce a perfect plane wave.

4.8. VELOCITY OF SOUND WAVES IN A GAS

Sound waves are mechanical waves which travel through the bulk of a solid, liquid, or gas. In a liquid or gas, only longitudinal waves are normally experienced. For as long as the wave length of the sound wave considerably exceeds the distance between the atoms of the substance through which it is travelling (and for sound waves in a gas, such as air, this is usually the case) the substance's atomic nature can be ignored in a simple study of the mechanics of the wave motion, and the substance can be treated as an elastic continuum.

Just as in the case considered in Section 4.6, where the equation of motion of a small section of the rope was written down and shown to have the form of the wave equation, the procedure to be followed in developing the theory of sound waves in a gas is to consider a small volume of the gas. Fig. 4.4 represents such a small volume, and one considers a plane wave propagating in the x direction. Before the wave passes, the undisturbed gas is represented by the region CDFE; after displacement, this time in the same direction as the wave is travelling, the region becomes C'D'F'E'. The plane through CD normal to the plane of the diagram has moved to become a parallel plane through C'D', a distance y away; similarly the other end of the region has been displaced a distance $\left\{y + \left(\frac{\partial y}{\partial x}\right)_t \delta x\right\}$.

Initially the gas is presumed to be at a pressure P. Accompanying the sound wave, there will be pressure changes, ΔP

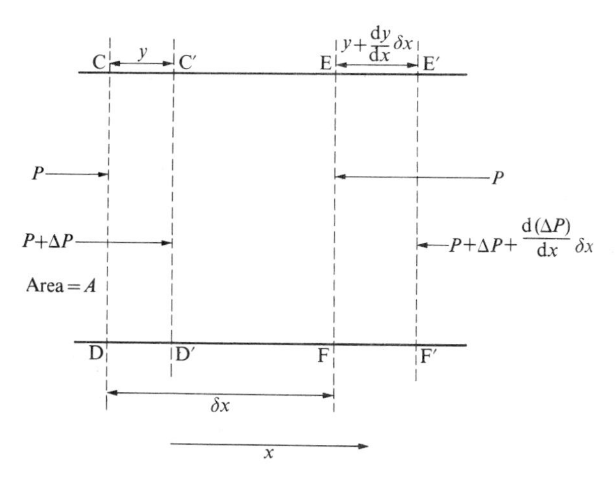

FIG. 4.4. A small volume of gas, CDFE, displaced to C'D'F'E' by the passage of a travelling wave in the x direction.

acting on the area A through C'D' and $\left\{\Delta P + \frac{\partial(\Delta P)}{\partial x}\delta x\right\}$ acting on the same area through E'F'. Let the density of the gas initially be ρ, so that the mass of gas in the volume under consideration is $\rho A\delta x$. The equation of motion (force = mass times acceleration) of the region in the x direction is therefore

$$(P + \Delta P)A - (P + \Delta P + \frac{\partial(\Delta P)}{\partial x}\delta x)A = \rho A\delta x\left(\frac{\partial^2 y}{\partial t^2}\right),$$

or

$$-\frac{\partial(\Delta P)}{\partial x} = \rho\left(\frac{\partial^2 y}{\partial t^2}\right). \qquad (4.17)$$

To proceed further, one needs to know more about ΔP. This comes from the elastic properties of the gas for one knows, by definition, that

$$K = -\frac{\Delta P}{\Delta V/V}$$

where K is the bulk modulus, V is the volume, and ΔV the volume change produced by the pressure change ΔP. This means

$$\Delta P = -K\frac{\Delta V}{V} . \qquad (4.18)$$

Inspection of Fig. 4.4 shows that $V = A\delta x$ and $\Delta V = A(\partial y/\partial x)\delta x$. Therefore

$$\Delta P = -K\left(\frac{\partial y}{\partial x}\right)$$

and eqn (4.17) becomes

$$-\frac{\partial}{\partial x}\left(-K\frac{y}{x}\right) = \rho\frac{\partial^2 y}{\partial t^2} .$$

This last equation can be rearranged in the form

$$\frac{\partial^2 y}{\partial t^2} = \frac{K}{\rho}\frac{\partial^2 y}{\partial x^2} . \qquad (4.19)$$

Eqn (4.19) can be recognized as the wave equation, eqn (4.12), for a wave whose velocity is given by

$$v = \left(\frac{K}{\rho}\right)^{\frac{1}{2}} . \qquad (4.20)$$

This result applies to any fluid where the only relevant elastic modulus is the bulk modulus K. The more general form

$$v = \left(\frac{\text{relevant elastic modulus}}{\text{density}}\right)^{\frac{1}{2}}$$

is useful for the case of solids where other moduli of elasticity may be needed to describe the distortion produced by the wave. To take however the specific case of a gas, to derive the velocity from eqn (4.20) it is necessary to know K. If a gas is compressed slowly, and is allowed to remain in thermal equilibrium with its surroundings, then the relevant bulk modulus is the isothermal bulk modulus; if however it is compressed quickly, thermal equilibrium will not be established and heat will remain in the gas. Conditions remain adiabatic, and the equation

$$PV^{\gamma} = \text{constant} \tag{4.21}$$

where γ is the ratio of the principal specific heats of the gas, will relate pressure and volume. Differentiating eqn (4.21), it follows that

$$P.\gamma V^{\gamma-1}\mathrm{d}V + V^{\gamma}\mathrm{d}P = 0,$$

which, on rearrangement gives

$$\gamma P = -\frac{\mathrm{d}P}{\mathrm{d}V/V} \tag{4.22}$$

This shows at once that the bulk modulus of the gas is γP, so that, in place of eqn (4.20), one can write

$$v = \left(\frac{\gamma P}{\rho}\right)^{\frac{1}{2}}. \tag{4.23}$$

This completes the calculation of the velocity of sound waves in a gas, provided all displacements are small; the result has, however, relevance to the study of sound in gases and to the estimation of value of γ for gases.

4.9. STANDING WAVES

One particular form of vibration, which could have been discussed in Chapter 3 but which is more appropriate to discuss here, is the so-called 'standing wave'. This is of such general importance and applicability that it is worthwhile discussing it briefly here.

One of the most easily demonstrable forms of standing wave is the vibration of a stretched string held firmly at both ends. Such a string could be the plucked string of a musical instrument. Fig. 4.5 illustrates possible forms of vibration. There is clear visual similarity with a sinusoidal wave motion; however, there is more than this, there is also theoretical justification for

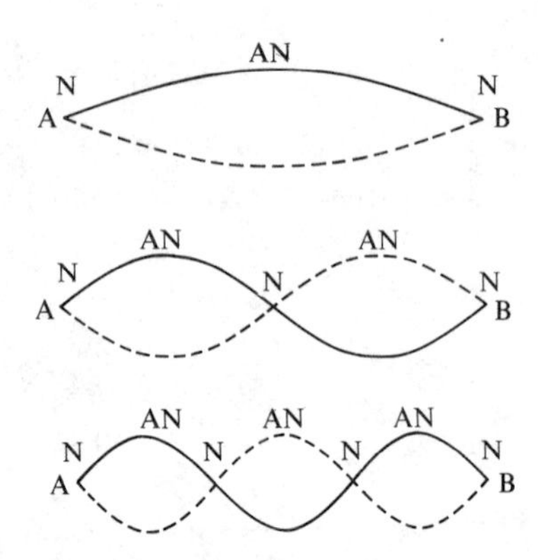

FIG. 4.5. Three possible forms of standing wave vibration on a string stretched between two fixed points A and B. The limits of the vibrations (much exaggerated relative to a real case) are indicated by the solid and the broken lines; points where the vibration amplitude is zero are called nodes (denoted by N) and places where the amplitude is a maximum are called antinodes (denoted by AN).

---

describing the motion as a form of wave motion.

To see this, consider two identical simple harmonic waves travelling in opposite directions down a rope. Let those two waves be described, as in eqn (4.2), by the equations

$$y_1 = a \sin \omega\left(t - \frac{x}{v}\right) \tag{4.24}$$

and

$$y_2 = a \sin \omega\left(t + \frac{x}{v}\right). \tag{4.25}$$

Eqn (4.24) represents a wave travelling in the $+x$ direction, whereas eqn (4.25) represents one travelling the other way. The overall displacement, y, at any point will follow from adding together y_1 and y_2, i.e.,

$$y = y_1 + y_2 = a\{\sin \omega\left(t - \frac{x}{v}\right) + \sin \omega\left(t + \frac{x}{v}\right)\}. \tag{4.26}$$

Use of the trigonometrical identity

$$\sin A + \sin B = 2 \cos\left(\frac{A - B}{2}\right) \sin\left(\frac{A + B}{2}\right)$$

allows one to rewrite eqn (4.26) in the form

$$y = 2a \cos \frac{\omega x}{v} \sin \omega t \qquad (4.27)$$

(it being permissible to drop the minus sign before $\omega x/v$). Comparison with eqn (3.3) makes it clear that the motion described by variations in y is SHM whose amplitude is a function of x, as it has the value $2a \cos (\omega x/v)$. As $(\omega x/v) = (2\pi x/\lambda)$, it becomes clear that the distance along the x direction between places where the amplitude is zero (the nodes) is $(\lambda/2)$.

Standing waves occur not only on strings, but also in the air columns of musical instruments, in the thermal vibrations of solids, and in the 'stationary states' encountered in the quantum mechanics of atomic phenomena. They represent a wave motion where it seems as though the wave is standing still; however, the theory above shows that such a motion can be regarded as the sum of two oppositely directed travelling waves, and they can arise if a travelling wave is reflected by a boundary, e.g., by the fixed end of a long rope.

4.10. ATTENUATION OF WAVES

It was shown in Chapter 3, eqn (3.20), that if a particle of mass m was executing SHM of amplitude a and angular frequency ω, then its energy was

$$E = \tfrac{1}{2}m\omega^2 a^2. \qquad (4.28)$$

This result allows one readily to obtain an expression for the intensity (Section 4.4) of a plane sound wave of amplitude a, angular frequency ω, travelling in a medium of density ρ with velocity v.

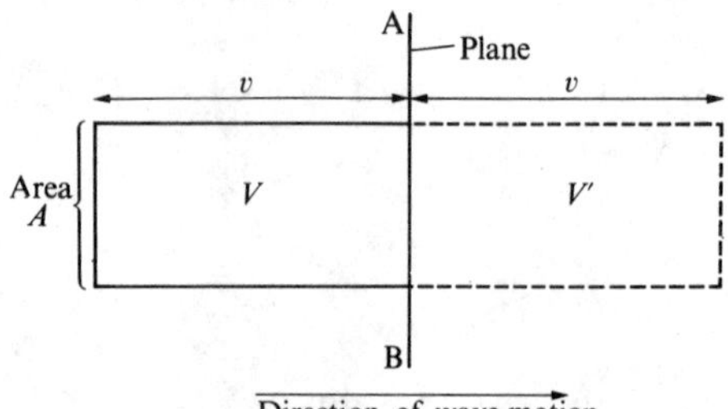

FIG. 4.6. To illustrate energy transfer by a wave motion.

In Fig. 4.6, AB represents a plane (normal to the diagram) through which a wave is passing in the direction shown. Consider the volume V shown; this is a cylinder of length v and area of cross-section A. The energy E contained in V as the sound wave passes is that of all the particles or all the mass in V executing SHM. That is to say,

$$E = \tfrac{1}{2}\,\rho v A \omega^2 a^2 , \tag{4.29}$$

by comparison with eqn (4.28) above.

Now after unit time, all the energy will have moved a distance v to the right, i.e. from V into V'. Therefore the intensity I of the wave (energy per unit area per unit time) is given by

$$I = \tfrac{1}{2}\,\rho v \omega^2 a^2 . \tag{4.30}$$

As with SHM, one notes that the energy (or intensity) is proportional to the square of the amplitude.

Applied to a wave motion, the term attenuation means the reduction of the intensity or amplitude as the wave progresses. When sound is radiated from a point source, or circular wave fronts on a water surface radiate out from the point where a stone has fallen in, there is only so much energy in the wave motion and this is clearly being dispersed over a greater area or distance the further the wave gets from the source; under such conditions, attenuation involves no absorption or scattering process. However, in practice, a plane wave can be attenuated, which means that energy is being lost from the wave as it progresses. It turns out that this rate of loss of energy is normally a constant fraction of the energy already there and hence is an exponential process. Consider then a plane wave whose amplitude is decreasing exponentially; substituting $y_o \exp(-\alpha x)$ for a in eqn (4.2) produces the equation

$$y_x = y_o \exp(-\alpha x) \sin \omega \left(t - \frac{x}{v}\right), \tag{4.31}$$

where α is defined to be the amplitude attenuation coefficient, If the amplitude has the values y_1 at $x = x_1$ and y_2 at $x = x_2$, it follows at once that

$$\alpha = \frac{1}{x_2 - x_1} \ln \frac{y_1}{y_2}.$$

The dimensions of α are clearly 1 length; the normal unit is simply m^{-1} (or, sometimes, nepers m^{-1}, as the logarithm of the ratio of the amplitudes is said to be in nepers). As the intensity is proportional to the square of the amplitude, the intensity attenuation coefficient is simply 2α.

The use of the exponential form of the equation for the wave [eqn (4.8) in place of eqn (4.2) in deriving eqn (4.31)] would have produced the equation

$$y_x = y_o \exp(-\alpha x) \exp i(\omega t - kx).$$

This can be rewritten as

$$y_x = y_o \exp i\{\omega t - (k - i\alpha)x\}.$$

indicating that attenuation can imply, from the theoretical point of view, a complex wave vector (i.e., $k - i\alpha$ instead of k) or, alternatively, a complex velocity.

The mechanism whereby plane waves are attenuated can vary very considerably from one type of wave to another. Often, especially in the case of sound waves, the energy appears as heat in the material. Sometimes however the energy can be scattered; perhaps the best known example of scattering is that of light where tiny particles in the atmosphere scatter blue light in preference to red, giving the sun a red appearance at sunset or through fog.

4.11. DISPERSION AND REFRACTION OF WAVES

In the examples of wave motion described in Sections 4.6 and 4.8, the velocity of the wave proved to be independent of the frequency. For audible sound waves in air, one knows this is true; a military band, heard from a distance, appears to be playing the same tune as if it were close to, so the sound from each of the different notes of different frequencies travels at the same velocity. However, very often it is not true that the velocity is independent of frequency; it may well depend on frequency, and in which case the medium through which the wave is travelling is said to be dispersive. The dispersion so produced can be described by some dispersion relationship or law relating velocity to frequency.

With mechanical or sound waves, dispersion can show itself when one can no longer treat the medium as uniform. When sound wavelengths become so short that they are near atomic spacings in magnitude (in a solid, this means wavelengths of a few times 10^{-10}m), then the velocity can depend quite appreciably on the sound frequency. The dispersion law is often shown as an 'ω - k' curve; Fig. 4.7(b) shows the form for this particular case. For low frequencies (long wavelength, low k), the velocity [$=\omega/k$, see eqn (4.7)] does not depend on frequency and no dispersion arises; as the wavelength approaches twice the interatomic spacing ($k = 2\pi/\lambda, \lambda = 2a$, $k = \pi/a$), then the velocity can be seen to drop (ω/k is less). This represents a real case,

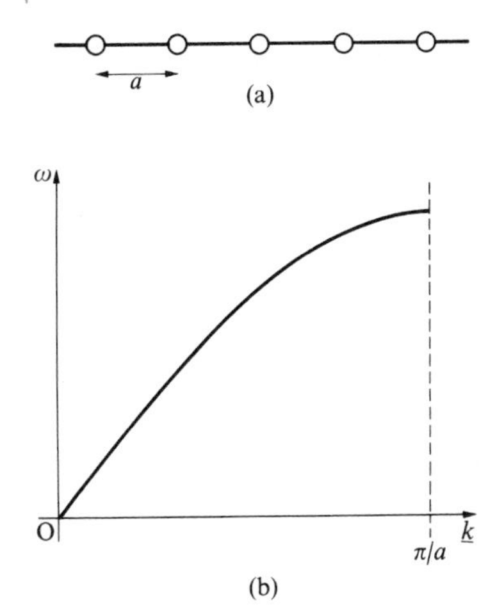

FIG. 4.7. Dispersion curve for mechanical waves passing along a chain of point masses separated by a distance a.

because thermal vibrations in a solid include such mechanical waves; the theory of specific heats, thermal conductivity etc., has to take account of this behaviour.

The association of velocity change at a boundary between two media with refraction of a wave is well known in the case of light waves, and the glass of a prism is clearly a dispersive medium for such waves. The same refraction phenomenon can occur with other types of waves, such as sound waves. The phenomenon becomes more complex when sound or light passes into a medium which is not isotropic, i.e. one whose properties can change with direction. The elastic modulus of a single crystal of many materials is very often a function of the direction of the applied stress. Under these conditions, refraction theory can become algebraically complex, even if not basically difficult, and phenomena such as double refraction, etc., can be evident.

4.12. POLARIZATION OF A WAVE

The direction or mode of the vibrations in a wave can be

called the polarization of the wave. Sound waves in solids are more complex than light waves in this regard, for in a solid both transverse and longitudinal waves are possible. In passing, it is of interest to note that a reason for believing that the earth has a liquid core is that transverse sound waves (generated by earthquakes) will not travel through this core, whereas they travel through the solid outside. A liquid will not carry a transverse wave any distance because it has no appropriate elastic response and flows rather than shears.

The propagation of a transverse wave through an anisotropic material such as a crystalline solid can be complicated by the response of the material to the impressed vibration; the phenomenon of double refraction, first noticed in calcite, where a beam of light is split into two beams, polarized at right angles to one another, is in fact commonplace with both light and sound. In considering the properties of transverse waves, it is often necessary to take into account the direction of the transverse polarization.

4.13. GROUP AND PHASE VELOCITIES

The wave velocity so far discussed is known as the phase velocity. In a non-dispersive medium, it is independent of frequency. In such a medium, a complicated vibration propagates unchanged as each of its Fourier components (i.e., the harmonics which together produce the wave form, see Section 3.4) travels with the same velocity. Therefore, eqn (4.7), i.e., $v = \omega/k$, is adequate to describe the movement of the whole vibration.

In a dispersive medium, on the other hand, the different Fourier components travel at different speeds. A complicated vibration can thus change its form as it progresses. This point can be illustrated by a very simple example shown in Fig. 4.8. Two waves, wavelengths λ and $\lambda + d\lambda$, are travelling together with velocities v and $v + dv$ through the same medium. The resulting disturbance of the medium is the sum of the disturbance caused by the two waves. Imagine the situation experienced by an observer travelling along at point A on the first wave. Fig. 4.8 (a) and (b) shows the situation when both waves produce a

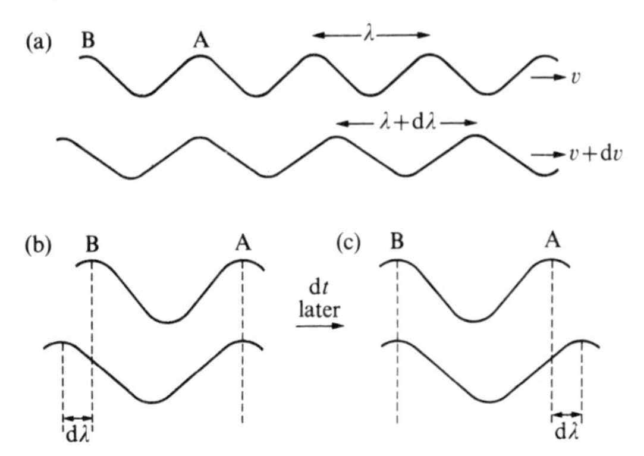

FIG. 4.8. To illustrate group movement; (a) and (b) show the two waves coinciding at A to give maximum displacement, (c) shows them dt later, when the maximum displacement has moved back to B.

maximum displacement at point A. The resulting wave form thus has its maximum at point A. Consider now a short time dt later, when the second wave has just advanced a distance $d\lambda$ relative to the first. This is shown in Fig. 4.8(c). The two waves are now producing their maximum displacement at point B. From the point of view of the observer travelling (with velocity v) at point A, the maximum in the resulting wave form has moved back to point B, a distance λ behind.

The complicated wave form composed of a number of Fourier components is called a wave group and its apparent velocity, in this example, the velocity of this maximum displacement, is called the group velocity. For the case given, this can be calculated in the following way. In time dt, the second wave has advanced on the first a distance $d\lambda$ at a relative velocity dv. Therefore

$$dt = \frac{d\lambda}{dv} . \tag{4.32}$$

Let the group velocity be u. Then in the same time dt, the group maximum has moved back (relative to point A) a distance λ at an apparent velocity $v - u$. (A purist might care to say moved a distance $-\lambda$ at a relative velocity $u - v$.) Thus it follows that

$$dt = \frac{\lambda}{v - u}. \tag{4.33}$$

Combining eqns (4.32) and (4.33) to eliminate dt, and rearranging it follows that the group velocity, u, is given by

$$u = v - \lambda \frac{dv}{d\lambda}. \tag{4.34}$$

It is a simple exercise to derive from this the alternative equation

$$u = \frac{d\omega}{dk}. \tag{4.35}$$

For practical purposes, group velocity is often more important than phase velocity, because the group velocity is the velocity with which energy is transferred by the wave motion. In the example given, this can readily be seen. Most of the energy lies where the displacement amplitude is at a maximum, and it is the velocity of this maximum, in amplitude and energy, which has been calculated as the group velocity. In the case where a dispersion curve of the $\omega - k$ type is available, the group velocity can be seen simply to be the slope of the curve; one can see in the example of Fig. 4.7 that the group velocity tends to zero as $k \to \pi/a$, a result which proves to be physically relevant as, for example, in the theory of solids, when treating vibrations of the crystal lattice.

The group velocity concept proves to be important in quantum mechanics in that the wave-like properties of a moving particle (such as an electron) are such that the particle proves to be moving with the group velocity of its guiding waves. This is not surprising as this is the velocity of energy transfer.

4.14. SURFACE WAVES

The first type of wave motion mentioned in this chapter, i.e., ripples on the surface of a pond, represents an example of a class of wave motions known as surface waves. The vibrations associated with them occur only close to the surface of the pond and die out rapidly in amplitude as one goes below the surface. The progation direction is along the surface, not into the medium. The vibrations are complex, a longitudinal and a transverse vibration taking place simultaneously so that the water is travelling in the direction of the wave at the wave crest and in the opposite direction at the trough. One can observe this in the behaviour of a floating object as sea waves pass it. Such surface waves normally show dispersion.

Surface waves also occur on solid surfaces. An example of this is provided by the waves produced on the surface of the earth by an earthquake. One of the types of waves produced in this way is known as the Rayleigh wave; Fig. 4.9 illustrates the characteristics of this form of wave. Recently, such surface waves have been produced on solids at frequencies of the order of a few hundred MHz, and it is possible that soon every colour television set will incorporate a filtering device using such surface waves. They are therefore of more than academic importance.

4.15. CONCLUSION

In this chapter, the main features of the mechanics of wave motion have been given a brief airing. The subject is a wide one because, in nature, there are many types of wave motion. The topic has been extensively studied and there is a great deal of literature on it; despite this, studies of current research

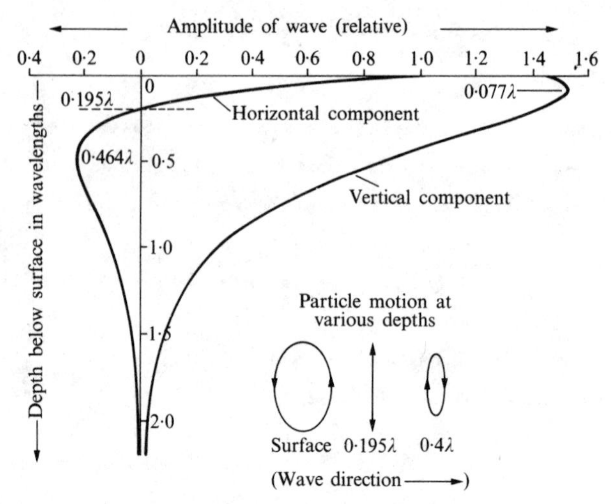

FIG. 4.9. To illustrate a (Rayleigh) surface wave. These curves were calculated from figures given in Rayleigh's original paper appropriate to a Poisson's ratio of 0.25.

publications show that the subject is far from exhausted and still offers many interesting problems.

4.16. EXAMPLES

1. A travelling wave is described by the equation

$$y_x = a \sin 2\pi \left(\frac{x}{\lambda} - ft\right).$$

Does this wave travel in the same direction as the wave described by the equation

$$y'_x = a \sin \omega \left(t - \frac{x}{v}\right)?$$

What is the relationship between y'_x and y_x?

2. Show that the velocity v of longitudinal sound waves travelling down a thin rod of material of density ρ and Young's modulus E is given by

$$v = (E/\rho)^{\frac{1}{2}}.$$

3. A steel wire of mass per unit length 0·00612 kg m^{-1} is stretched between two fixed points a distance 1·00 m apart. What tension must it be under if, when plucked, it vibrates at the note A, i.e., at a frequency of 440 Hz?

4. The velocity of sound in air at normal atmospheric pressure and at 20°C is 344 m s^{-1}. What is the velocity of sound in helium at the same pressure and temperature? (For air, density at 20°C = 1·29 g/l. and γ = 1·40; for helium, density at 20°C = 0·178 g/l. and γ = 1·67.)

5. A plane sound wave is being attenuated exponentially. After travelling 5·00 m, its amplitude is halved. What is the attenuation coefficient in nepers m^{-1} ? The intensity attenuation coefficient is often expressed in decibels (dB)m^{-1}; by definition, the difference in intensities in dB is ten times the logarithm to base 10 of the ratio of the intensities. For this particular sound wave, what is the attenuation coefficient in dB m^{-1} ?

6. The phase velocity v of a wave motion is found to depend on the wavelength λ according to the equation

$$v = a\lambda^{x},$$

where a and x are constant. Show that the group velocity u for this wave motion is given by

$$u = (1 - x)v.$$

Waves on deep water move with a phase velocity v given by $v^2 = g\lambda/2\pi$ where g is the acceleration due to gravity. Show that the group velocity for these waves is one half of the phase velocity.

7. de Broglie's hypothesis states that a particle of mass m travelling with velocity v has wave-like properties, and the

associated waves (known as de Broglie waves) have a wavelength λ given by the equation;

$$\lambda = \frac{h}{mv}$$

Substituting $\hbar$ for $h/2\pi$, this expression is equivalent to writing for the momentum p of the particle the equation

$$p = \hbar k,$$

where $k = 2\pi/\lambda$. The angular frequency ω of the waves is assumed to be related to the total energy E of the particle by

$$E = \hbar\omega.$$

Furthermore, the special theory of relativity provides the equations

$$E = mc^2 = (c^2p^2 + m_o^2c^4)^{\frac{1}{2}},$$

where m_o is the rest mass of the particle and c is the velocity of light. Calculate (a) the phase velocity, and (b) the group velocity of the de Broglie waves.

5. Motion under a central force

5.1. MEANING OF A CENTRAL FORCE

The motion illustrated in Fig. 2.5 and discussed in Section 2.3., namely a mass moving in a circular orbit under the influence of a centripetal force, is an example of motion under the action of a central force. If one whirls a mass on the end of a string around in a circle, holding the string in one's hand, then the relevant force on the mass is always directed centrally along the line of the string towards one's hand. It is a central force. Central forces are those which are always directed towards one particular point in space, regardless of where the mass experiencing the force is located.

Such forces do not need a string to produce them. The planets move in their orbits under the action of a central force, i.e., the gravitational pull of the sun, always directed towards the sun. The electrostatic field of a point electric charge can give rise to a central force on any other electric charge. In both cases, the force acts at a distance through space. (The physicist's model for the mechanism is the concept of a field of force). Motion under the action of such a force can be important and will now be discussed.

5.2. PLANETARY MOTION AND KEPLER'S LAWS

Perhaps the best known example of motion under the action of a central force is the case, already mentioned, of the planets moving in orbits around the sun. The nature of planetary orbits was only established shortly before Newton was to develop his theories of mechanics and of gravitation. In 1609, the German astronomer Kepler, analysing the observations of the Danish astronomer Tycho Brahe (to whom he had been an assistant), concluded that the planet Mars behaves as follows:

(1) The planet moves in an ellipse which has the sun as one of its foci;

(2) the line joining the sun to the planet sweeps out equal areas in equal times.

After further analysis, Kepler, in 1618, reported that these two laws applied to the other planets and added a third law:

(3) The square of the period of the orbit of any planet is proportional to the cube of its distance from the sun.

These three laws are known as Kepler's laws of planetary motion.

5.2.1. *Kepler's second law*

It turns out that Kepler's second law is general to any particle travelling under the action of a single central force only irrespective of the law of force, i.e., how the force depends on the particle's distance from its source. It turns out to be just a statement of the conservation of angular momentum. Consider a particle of mass m moving with velocity $\underline{v}$ a distance $\underline{r}$ from the source of a central force, as illustrated by Fig. 5.1, where

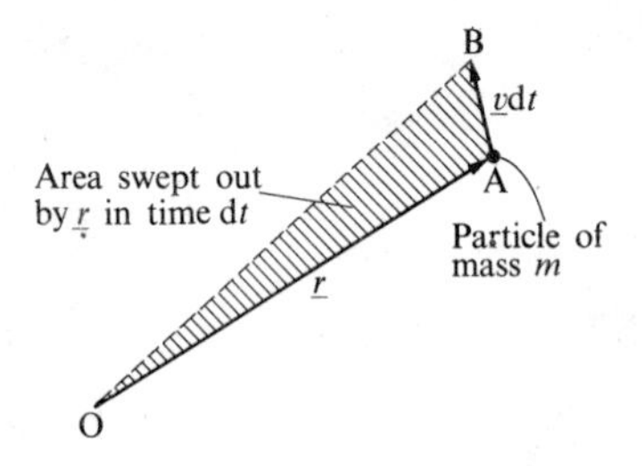

FIG. 5.1. A particle of mass m moves from A to B in time dt; the area swept out by the radius vector $\underline{r}$ is that of the triangle OAB.

both $\underline{r}$ and $\underline{v}$ lie in the plane of the diagram. The angular momentum of the particle about O, $\underline{L}$, is given as in eqn (2.25), by

$$\underline{L} = \underline{r} \times m\underline{v}$$

$$= m(\underline{r} \times \underline{v}).$$

The central force, acting towards O, can have no effect on the angular momentum; as no other forces are acting, it follows that $\underline{L}$ is constant, so, throughout the motion

$$(\underline{r} \times \underline{v}) = \text{constant} \qquad (5.1)$$

Inspection of Fig. 5.1 shows that the element of area, ds, swept out by r in time dt is given by

$$\mathrm{d}\underline{s} = \tfrac{1}{2}(\underline{r} \times \underline{v}\ \mathrm{d}\underline{t}),$$

so that the rate at which the area is being swept out is given by

$$\frac{\mathrm{d}\underline{s}}{\mathrm{d}t} = \tfrac{1}{2}\ \underline{r} \times \underline{v}$$

which, from eqn (5.1), also equals a constant.

Thus it follows that Kepler's second law is a natural consequence of the conservation of angular momentum.

One other consequence of the constancy of $\underline{L}$ is that $\underline{L}$ does not move in space; as $\underline{L}$ is always at right angles both to $\underline{r}$ and to $\underline{v}$, it follows that the path of any mass m under the action of a central force lies in a plane through the point of the origin of the force. That the orbits of planets are planar, and that the sun lies in this plane, is thus also accounted for by Newton's mechanics.

5.2.2. *Newton's law of gravitation and gravitational potential energy*

In 1666, Isaac Newton, an undergraduate at Cambridge, was sent home when the university was closed as a result of the plague; during this enforced vacation, he saw that Kepler's third law implied that the same gravitational force, which could cause an apple to fall from a tree, could provide the force needed to keep the planets in their orbit provided that it was an inverse square law of force. This means that two masses, M and m, a distance r apart, will attract one another with a force $\underline{F}$ given (in scalar notation) by

$$F = G\frac{Mm}{r^2}, \tag{5.2}$$

where G is a constant (now known as the gravitational constant). The inverse square applies to the distance r. Eqn (5.2) is known as Newton's law of gravitation; it does not tell one what gravity is, it merely describes how gravity behaves.

The potential energy of a particle in a gravitational field due to a mass M can be found as follows. Let a particle of mass m be brought from infinity to a distance r away from the fixed mass M. Let the potential energy of m at infinity be zero. Then, when m is at r, the work that can be obtained from it by letting it return to infinity (i.e., its PE) will be

$$-\int_r^\infty F\mathrm{d}r = \int_r^\infty - G\frac{Mm}{r^2}\,\mathrm{d}r = -G\frac{Mm}{r}. \tag{5.3}$$

So the potential energy will (not surprisingly) be negative. (The gravitational potential at the point will be the potential energy per unit mass, i.e. $-GM/r$).

5.2.3. *Motion under an inverse square law of force*

The inverse square law, as described by Newton's law of gravitation, can be applied to the problem of planetary motion as follows. Consider a small mass m moving under the influence of a gravitational force from a large mass M, which can be treated as stationary. Let M be at the origin of polar coordinates r and θ, so that m is at the point r,θ. Then, from Fig.5.2, the rate at which the line joining m and M is sweeping out area is given by $\frac{1}{2}r^2(\mathrm{d}\theta/\mathrm{d}t)$; this is a constant (Section 5.2.1), A say, i.e.

$$\tfrac{1}{2}r^2\left(\frac{\mathrm{d}\theta}{\mathrm{d}t}\right) = A. \qquad (5.4)$$

Throughout the orbit, the kinetic energy + gravitational potential energy of m must remain constant as no other forces are acting, so, using eqn (5.3),

$$\tfrac{1}{2}mv^2 - G\frac{Mm}{r} = B, \qquad (5.5)$$

where B is some other constant (the total energy of m).

To express v in terms of polar coordinates, one can use the triangle of Fig. 5.3, and note that

$$v^2\mathrm{d}t^2 = \mathrm{d}r^2 + r^2\mathrm{d}\theta^2,$$

so that eqn (5.5) can be rewritten as

$$\tfrac{1}{2}m\left\{\left(\frac{\mathrm{d}r}{\mathrm{d}t}\right)^2 + r^2\left(\frac{\mathrm{d}\theta}{\mathrm{d}t}\right)^2\right\} - G\frac{Mm}{r} = B. \qquad (5.6)$$

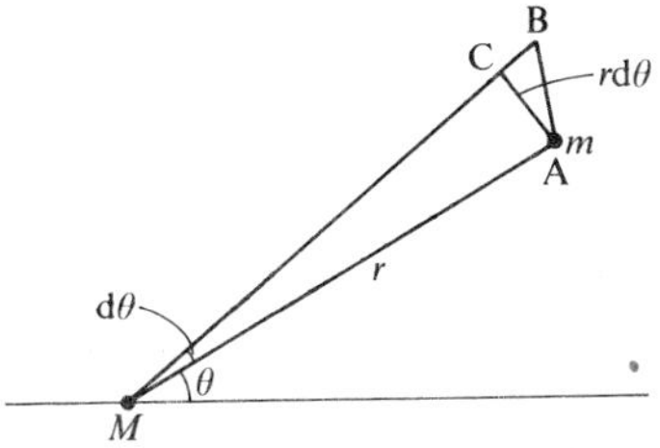

FIG. 5.2. Fig. 5.1 redrawn in polar coordinates. Area swept out in time $\mathrm{d}t = \frac{1}{2}r^2\mathrm{d}\theta$.

Noting that

$$\frac{\mathrm{d}r}{\mathrm{d}t} = \frac{\mathrm{d}r}{\mathrm{d}\theta}\,\frac{\mathrm{d}\theta}{\mathrm{d}t},$$

eqn (5.6) becomes

$$\tfrac{1}{2}m\left\{\left(\frac{\mathrm{d}r}{\mathrm{d}\theta}\right)^2\left(\frac{\mathrm{d}\theta}{\mathrm{d}t}\right)^2 + r^2\left(\frac{\mathrm{d}\theta}{\mathrm{d}t}\right)^2\right\} - G\frac{Mm}{r} = B. \qquad (5.7)$$

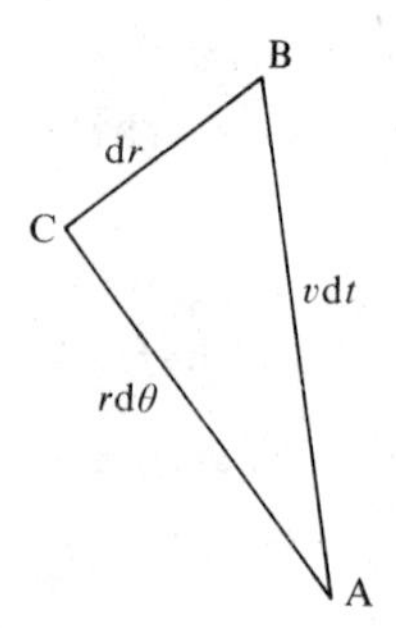

FIG. 5.3. An enlargement of part of Fig. 5.1. and 5.2.

Referring back to eqn (5.4), it becomes possible with its help to substitute $(2A/r^2)$ for $(d\theta/dt)$, so that one obtains

$$\tfrac{1}{2}m\left\{\frac{4A^2}{r^4}\left(\frac{dr}{d\theta}\right)^2+\frac{4A^2}{r^2}\right\}-G\frac{Mm}{r}=B. \qquad (5.8)$$

The time has been eliminated so that eqn (5.8) is a differential equation relating the two coordinates r and θ describing the position of m in the orbit. Any possible orbit of m must be a solution of this equation. Dividing through by $\tfrac{1}{2}m$ and moving all terms not containing θ to the right-hand side, the equation becomes

$$\frac{4A^2}{r^4}\left(\frac{dr}{d\theta}\right)^2=\frac{2B}{m}+\frac{2GM}{r}-\frac{4A^2}{r^2}$$

or

$$\frac{1}{r^4}\left(\frac{dr}{d\theta}\right)^2=\frac{B}{2A^2m}+\frac{GM}{2A^2r}-\frac{1}{r^2}. \qquad (5.9)$$

Taking square roots and rearranging so as to separate the variables r and θ, one gets

$$d\theta=\frac{dr}{r^2}\Big/\left(\frac{B}{2A^2m}+\frac{GM}{2A^2r}-\frac{1}{r^2}\right)^{\frac{1}{2}} \qquad (5.10)$$

The equation can now be integrated; this procedure is simplified by the substitution $u = 1/r$, so that $du = -dr/r^2$, and the equation becomes

$$\int d\theta = -\int du \Big/ \left(\frac{B}{2A^2 m} + \frac{GM}{2A^2} u - u^2 \right)^{\frac{1}{2}} \tag{5.11}$$

The right-hand side is an integral of standard form; the solution to

$$\int dx \Big/ \left(ax^2 + 2bx + c \right)^{\frac{1}{2}} = \left(-a \right)^{\frac{1}{2}} \cos^{-1} \left\{ (ax + b)/(b^2 - ac)^{\frac{1}{2}} \right\} ,$$

so, applying this result to the right-hand side of eqn (5.11), at the same time writing c for $(B/2A^2 m)$ and b for $(GM/2A^2)$ (for convenience), one obtains

$$\theta + c = -\cos^{-1} \left\{ (b - u)/(b^2 - c)^{\frac{1}{2}} \right\}$$

where C is a constant of integration. Taking cosines, noting that $\cos(-\phi) = \cos\phi$, it follows that

$$\cos(\theta + c) = (b - u)/(b^2 - c)^{\frac{1}{2}}$$

or

$$u = b - (b^2 - c)^{\frac{1}{2}} \cos(\theta + C).$$

Replacing u by $1/r$, this means

$$r = \frac{1/b}{1 - (1 - c/b^2)^{\frac{1}{2}} \cos(\theta + C)} ,$$

i.e.,

$$r = \frac{2A^2/GM}{1 - (1 - 2A^2 B/G^2 M^2 m) \cos(\theta + c)} . \tag{5.12}$$

Now this takes the form of a polar equation of a conic section, either an ellipse, parabola, or hyperboloid. The polar equation for an ellipse where r is measured from a focus can be written as

$$r = \frac{b/a}{1 - e\cos\theta} , \tag{5.13}$$

where the ellipse has major and minor axes (diameters) $2a$ and $2b$ respectively and e is the eccentricity $\left[= \left\{(a^2 - b^2)/a^2\right\}\right]$ which is > 1. If $e = 1$, the equation represents a parabola and, if $e < 1$, a hyperbola. The constant C in eqn (5.12) fixes the axes and can clearly be taken as zero for discussion purposes. Now

$$e = 1 - 2A^2B/G^2M^2m,$$

so $e > 1$ must mean that B is negative; reference to eqn (5.5) shows that this means the total energy is negative, i.e., the magnitude of the negative potential energy exceeds the kinetic energy. The conditions for parabolic and hyperbolic orbits are also clear.

The only repeating or closed orbit is one of elliptical form. Had the law of force not been that of an inverse square, then the potential energy term to be included in eqn (5.5) would not have been $-GMm/r$ and the orbit would no longer have been of the elliptical form described. The result is that Newton's law of gravitation accounts for the first of Kepler's laws.

In passing, it is worth noting that, if the law of force were repulsive rather than attractive, then an elliptical orbit would not be possible as the total energy would always be positive. This result is of course only common-sense and hardly needs formal proof.

Kepler's third law also follows readily from these results. Let the period of a planet in its orbit be T. As the area of the orbit is πab, it follows from the definition of A as the rate at which area is being swept out that

$$AT = \pi ab$$

or

$$T \propto \frac{ab}{A} \tag{5.14}$$

Eqns (5.12) and (5.13) give one, by inspection, the equation

$$\frac{b^2}{a} = \frac{2A^2}{GM}$$

i.e.

$$b \propto (A^2 a)^{\frac{1}{2}}. \tag{5.15}$$

Substituting for b from eqn (5.15) into eqn (5.14) gives at once

$$T \propto \frac{a}{A} (A^2 a)^{\frac{1}{2}}$$

or

$$T^2 \propto a^3, \tag{5.16}$$

which is substantially a statement of Kepler's third law; the eccentricity of planetary orbits is small.

It is important to note that throughout this argument, it has been assumed that the source of the force, the sun, can be treated as a stationary object. A strict calculation would not permit this; one should treat as stationary the centre of mass of sun and planet, which can only approximate in position to the centre of mass of the sun; the approximation is adequate for the purpose of this discussion, but not for strict astronomical work. Similarly, in such work, the gravitational forces between planets must be taken into account. The discovery of the planet Neptune in 1846 followed such calculations by the mathematicians J.C. Adams and J.J. Leverrier, who independently predicted its existence from the observed motions of the other planets and so led observers to find it; the planet Pluto was found by similar means in 1930.

5.3. USE OF POLAR COORDINATES TO DESCRIBE MOTION IN A PLANE

The last section has made it clear that polar coordinates are extremely useful when dealing with central force problems.

It is therefore worthwhile to digress at this point, and to discuss these coordinates in a little more detail, incorporating the advantages of vector notation into the discussion.

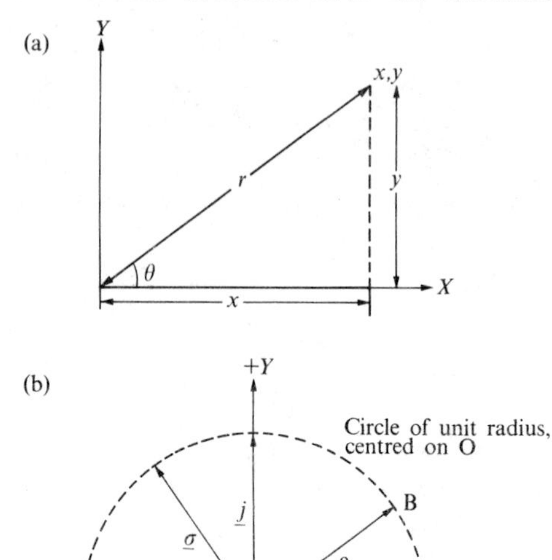

FIG. 5.4. (a) Polar coordinates; (b) Unit vectors.

Simple polar coordinates are illustrated in Fig. 5.4(a). A point can be described either by the cartesian coordinates x, y, or the polar coordinates r,θ. In scalar notation, one can write at once

$$x = r\cos\theta, \tag{5.17}$$

$$y = r\sin\theta. \tag{5.18}$$

Now let $\underline{i},\underline{j}$ be unit vectors along the X and Y axes. Then clearly

$$\underline{x} = \underline{i}r\cos\theta, \tag{5.19}$$

$$\underline{y} = \underline{j}r\sin\theta; \tag{5.20}$$

it is furthermore obvious from this that

$$\underline{r} = \underline{x} + \underline{y} = r(\underline{i}\cos\theta + \underline{j}\sin\theta). \tag{5.21}$$

Let $\underline{\rho}$ be a unit vector in the direction of $\underline{r}$, so that

$$\underline{r} = \underline{\rho}\ r. \tag{5.22}$$

Eqns (5.21) and (5.22) show at once that

$$\underline{\rho} = \underline{i} \cos\theta + \underline{j} \sin\theta. \tag{5.23}$$

This result follows also from inspection of Fig. 5.4(b). Inspection of this figure also shows that, if $\underline{\sigma}$ is a unit vector directed at right angles to $\underline{r}$ in the direction shown, then

$$\underline{\sigma} = \underline{j} \cos\theta - \underline{i} \sin\theta. \tag{5.24}$$

The two quantities of most interest are the velocity ($\underline{v}$) and acceleration (a). To find the first of these in polar coordinates, one first differentiates eqns (5.19) and (5.20) with respect to time and obtains

$$\dot{\underline{x}} = \underline{i}(\dot{r} \cos\theta - r\,\dot{\theta} \sin\theta), \tag{5.25}$$

$$\dot{\underline{y}} = \underline{j}(\dot{r} \sin\theta + r\,\dot{\theta} \cos\theta), \tag{5.26}$$

whence $\underline{v} = \dot{\underline{x}} + \dot{\underline{y}} = r(\underline{i} \cos\theta + \underline{j} \sin\theta) + r\dot{\theta}(\underline{j} \cos\theta - \underline{i} \sin\theta),$ (5.27)

or, from eqns (5.23) and (5.24)

$$\underline{v} = \underline{\rho}\dot{r} + \underline{\sigma}r\dot{\theta} \tag{5.28}$$

This tells one that the velocity has a component $\dot{r}$ in the direction of $\underline{r}$ and a component $r\dot{\theta}$ perpendicular to it; the magnitude of $\underline{v}$ follows at once in the usual way, as clearly

$$v^2 = \dot{r}^2 + r^2\dot{\theta}^2 \tag{5.29}$$

giving

$$v = (\dot{r}^2 + r^2\dot{\theta}^2)^{\frac{1}{2}}. \tag{5.30}$$

Eqn (5.29) has of course already been derived and used in obtaining eqn (5.6).

The acceleration is slightly more complicated and less obvious, although the result will be seen to be easily understood. Differentiating again w.r.t. time, eqns (5.25) and (5.26) become

$$\ddot{\underline{x}} = \underline{i}(\ddot{r}\cos\theta - \dot{r}\sin\theta\dot{\theta} - \dot{r}\sin\theta\dot{\theta} - r\cos\theta\dot{\theta}^2 - r\sin\theta\ddot{\theta})$$

and $\ddot{y} = \underline{j}\,(\dot{\underline{r}}\sin\theta + \dot{r}\cos\theta\dot{\theta} + \dot{r}\cos\theta\dot{\theta} - \underline{r}\sin\theta\dot{\theta}^2 + \underline{r}\cos\theta\ddot{\theta})$,

i.e., $$\ddot{\underline{x}} = \underline{i}\,\{(\ddot{r} - r\dot{\theta}^2)\cos\theta - (2\dot{r}\dot{\theta} + r\ddot{\theta})\sin\theta\} \quad 5.31)$$

and $$\ddot{\underline{y}} = \underline{j}\,\{(\ddot{r} - r\dot{\theta}^2)\sin\theta + (2\dot{r}\dot{\theta} + r\ddot{\theta})\cos\theta\} \quad (5.32)$$

Using the fact that $\underline{a} = \ddot{\underline{x}} + \ddot{\underline{y}}$, and incorporating the results of eqns (5.23) and (5.24), it follows that

$$\underline{a} = \underline{\rho}\,(\dot{r} - r\dot{\theta}^2) + \underline{\sigma}\,(2\dot{r}\dot{\theta} + \ddot{\theta}). \quad (5.33)$$

This result is, as already stated, easily understood. There are four terms; the first two, directed along $\underline{r}$, are acceleration due to acceleration of $\underline{r}$ outwards and centripetal acceleration inwards, while the second two, perpendicular to $\underline{r}$, consist of the Coriolis acceleration and the consequence of angular acceleration respectively.

Eqn (5.33) is general. For the particular case of the central force problem, there is no acceleration component perpendicular to $\underline{r}$; this can most easily be derived by differentiation of eqn (5.4) which soon leads to the equation

$$2\dot{r}\dot{\theta} + r\ddot{\theta} = 0, \quad (5.34)$$

but it can also be seen to follow from the fact that any acceleration of a mass in this direction implies a torque and an associated change in angular momentum, neither of which exist. Therefore, for the particular case of the central force problem,

$$a = \ddot{r} - r\dot{\theta}^2. \quad (5.35)$$

This result could easily have been derived by an argument omitting unit vectors but otherwise similar. The directions of the components of velocity and acceleration would not however have been made clear and the picture would thereby have been incomplete.

5.4. GENERAL DIFFERENTIAL EQUATIONS FOR AN ORBIT UNDER THE ACTION OF A CENTRAL FORCE

Consider a particle moving under a central force. Whatever the nature of the force, eqn (5.4) will apply, i.e.

$$\tfrac{1}{2}r^2\dot{\theta} = A$$

or

$$\dot{\theta} = \frac{2A}{r^2}. \tag{5.36}$$

Let the central force, F, be some function of the distance of the particle from the source of the force, i.e., let $F = f(r)$. It follows therefore that the equation of motion of the particle ($F = ma$) takes, using eqn (5.35), the form

$$f(r) = m(\ddot{r} - r\dot{\theta}^2). \tag{5.37}$$

It is worthwhile at this stage to introduce the substitution $u = 1/r$ which means that

$$\dot{\theta} = 2Au^2, \tag{5.38}$$

also

$$\dot{r} = \frac{dr}{d\theta}\,\dot{\theta} = \left(-\frac{1}{u^2}\frac{du}{d\theta}\right).2Au^2 = -2A\frac{du}{d\theta}, \tag{5.39}$$

and

$$\ddot{r} = -2A\frac{d}{dt}\left(\frac{du}{d\theta}\right) = -2A\frac{d}{d\theta}\left(\frac{du}{d\theta}\right)\dot{\theta} = -4A^2u^2\frac{d^2u}{d\theta^2}. \tag{5.40}$$

Substituting now from eqns (5.38), (5.39), and (5.40) into eqn (5.37), it follows that

$$f(r) = m\left(-4A^2u^2\frac{d^2u}{d\theta^2} - \frac{1}{u}\,4A^2u^4\right)$$

or

$$\frac{d^2u}{d\theta^2} + u - \frac{f(r)}{4A^2mu^2} = 0. \tag{5.41}$$

This is a second-order differential equation, the solution of which will give the orbit in terms of $u(= 1/r)$ and θ, and which will incorporate two constants of integration.

An alternative starting point was used in Section 5.2.3, in that the fact that the total energy remained constant was used, rather than the equation of motion, $F = ma$. If one assumes that the potential energy takes the form $V(u)$, then, for the general case, the equation to replace eqn (5.6) would take the form

$$\tfrac{1}{2}\,m(\dot{r}^2 + r^2 + r^2\dot{\theta}^2) + V(u) = B \tag{5.42}$$

where B is the constant total energy. Substituting now for r, $\dot{r}$ and $\dot{\theta}$ in the kinetic energy term of this equation, it can easily be rearranged in the form

$$\left(\frac{du}{d\theta}\right)^2 + u^2 = \frac{B - V(u)}{2A^2m} \tag{5.43}$$

Eqn (5.43) is a first-order differential equation and is a perfectly valid alternative to eqn (5.42). It will only have one constant of integration; the total energy, B, is a second extra constant, so no diminution in the number of constants has occurred. The integration constants are in any event set by the initial conditions. Eqn (5.43) is always integrable by the technique of separation of variables, but the solution, and ease of solution, will depend on the form of $V(u)$.

The fact that the total energy remains constant over an orbit, and that the potential energy is a single-valued function of position, means that no work has been done on or by the mass m during an orbit. Thus over any closed orbit

$\oint \underline{F} \cdot d\underline{l}$ = zero. This can only occur when the field of force is said to be conservative. There are fields of force which can be non-conservative (e.g., the magnetic field around a long straight wire carrying a current), but in the central force problem, one is essentially dealing only with a conservative force field.

5.5. SATELLITES AND THE BOHR ATOM

The study of planetary motion has, during recent years, received practical application in the use of artificial earth satellites in communications, weather forecasting, etc., and in the various aspects of space travel and space research. The calculation of satellite behaviour involves the central force problem; although in this case the force from the earth is not the only force acting, it dominates sufficiently for the results to remain approximately true. In the neighbourhood of the earth, there is the additional complication of friction due to the earth's atmosphere. If one ignores this for the moment, it is relatively easy to calculate the 'escape velocity', that is to say, the minimum velocity which one would have to give an object at the earth's surface to ensure that it would never return. This means that e from eqn (5.13), shall be $\leqslant 1$, i.e., that the total energy B from eqn (5.5), shall be zero or positive. The escape velocity will have $B = 0$. In other words, the object must have sufficient kinetic energy to take it to infinity against the earth's gravitational force.

Another application of planetary motion under the action of a central force was the model of the atom proposed by Bohr in 1913. In this model, the atom consists of a small heavy positively charged mass, the nucleus, around which lighter negatively charged masses, the electrons, move in circular orbits; the central force is the electrostatic (Coulomb) force obeying an inverse square law. The model incorporates quantum restrictions in that the angular momentum of each electron in its orbit is restricted to a value which is an integer times $\hbar$ where $\hbar$ is Planck's constant divided by 2π. This restriction brings in one effect of quantum theory on the theory of mechanics, but it is not an effect which can alter the fundamental principles and laws. Sommerfeld extended the model to include elliptical orbits subject to a similar quantum restriction. The model has since been replaced, but remains of interest.

If an electron is moving very fast in its orbit, the special theory of relativity predicts that its mass will change; Sommerfeld also introducted the effect of this to atomic theory.

In planetary orbits, the general theory of relativity also predicts effects; in particular, the theory has accounted for the slow rotation of the whole ellipse of the orbit of the planet Mercury.

5.6. REPULSIVE FORCES AND ATOMIC SCATTERING

Consider a small charged particle, of mass m, charge $+ze$, rapidly approaching a large stationary charged particle, of mass M, charge $+Ze$, where Z and z are integers and e is the magnitude of the electronic charge. Such a situation occured in one of the earliest important experiments in nuclear physics when, in 1911, Rutherford allowed a beam of fast-moving α-particles (ionized helium nuclei, carrying positive charge $2e$) to strike a thin gold foil and be scattered by the heavy gold nuclei (carrying positive charge $79e$ and of mass equal to nearly 50 α-particles). It is now well known that most α-particles went through the foil unscattered, leading Rutherford to estimate that the scatterer could only take up a very small part of the volume occupied by the atom and hence to suggest that this scatterer was a small heavy nucleus in the atom.

Consider now the theory of the scattering process, when the approaching particle is experiencing only the central repulsive force (F) from the scatterer. From Coulomb's law, the magnitude of this force will be given by

$$F = \frac{Zze^2}{4\pi\varepsilon_o r^2} \tag{5.44}$$

where r is the distance between the particles and ε_o is the permittivity of free space. For simplicity, it is convenient here to substitute X for $Zze^2/4\pi\varepsilon_o$ and to write

$$F = \frac{X}{r^2} \,. \tag{5.45}$$

A possible form of the path to be followed by the scattered particle is shown in Fig. 5.5. The form of this path can be

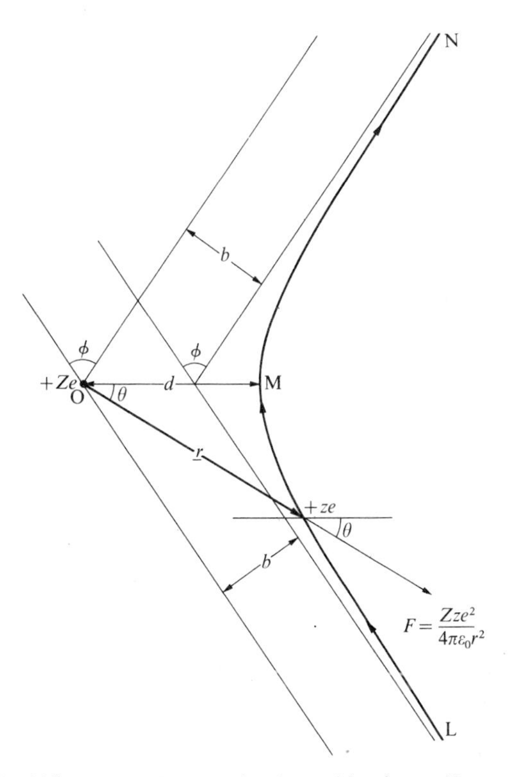

FIG. 5.5. To illustrate Rutherford scattering, the scattering particle being at O and the path followed by the scattered particle being LMN.

derived by the method used in Section 5.2.3, when knowledge of the geometry of the hyperbola would enable one to relate the angle of scattering (ϕ) to the impact parameter (b), This 'impact parameter' is the distance by which the approaching particle would have missed the scatterer had there been no scattering force. For most practical purposes, though, a detailed expression for the shape of the path is not required, and ϕ can be related to b in the following way.

Let the scattered particle have mass m and let it approach the scatterer from a great distance with speed v_o. While it will slow down along the part of the path LM as part of its kinetic energy becomes potential energy, it will speed up again along MN and at a great distance (potential energy → zero) will regain the speed v_o. The resulting change in momentum can be seen from the vector diagram of Fig. 5.6. to be

$$2mv_o \sin(\phi/2) \tag{5.46}$$

in the direction of OM on Fig. 5.5. This can be equated to the

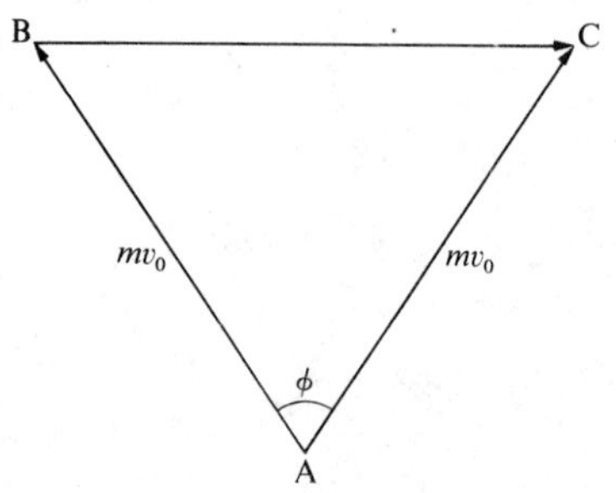

FIG. 5.6. To illustrate the momentum change during Rutherford scattering; AB = initial momentum, AC = final momentum, and BC = change of momentum.

impulse received by the particle during the path (see Section 1.7), so that, considering only the component of impulse in the direction OM (there is no need to consider the other components as no momentum change is involved for them), one can write

$$\int_{-\theta}^{\theta} F \cos \theta \, dt = 2mv_o \sin (\phi/2). \tag{5.47}$$

In all central force problems, angular momentum is conserved (Section 5.2.1). The angular momentum, using eqns (2.25) and (5.28), has a magnitude of

$$|\underline{r} \times m\underline{v}| = mr^2 \frac{d\theta}{dt} \tag{5.48}$$

and has an initial value of mv_ob. This conservation therefore means that, equating mv_ob to $mr^2(d\theta/dt)$,

$$\frac{dt}{d\theta} = \frac{r^2}{v_ob} \tag{5.49}$$

Now $\quad dt = \dfrac{dt}{d\theta} d\theta$ and $\therefore \quad = \dfrac{r^2 d\theta}{v_ob}$. (5.50)

Substituting from eqns (5.50) and (5.45) into eqn (5.47), it follows that

$$\int_{\theta_1}^{\theta_2} \frac{X}{r^2} \cos \theta \frac{r^2 d\theta}{v_ob} = 2mv_o \sin (\phi/2) \tag{5.51}$$

where $\theta_1 = -(90 - \phi/2)$ and $\theta_2 = (90 - \phi/2)$ (see Fig. 5.5).

Therefore, on integrating,

$$\frac{2X}{v_ob} \sin (90 - \phi/2) = 2mv_o \sin (\phi/2). \tag{5.52}$$

As $\sin (90 - \phi) = \cos \phi$, it follows that

$$\frac{2X}{v_ob} = 2mv_o \tan (\phi/2)$$

or

$$b = \frac{X}{mv_o^2} \cot (\phi/2). \tag{5.53}$$

Resubstituting for X, this becomes

$$b = \frac{Zze^2}{4\pi\varepsilon_o mv_o^2} \cot (\phi/2), \tag{5.54}$$

and this is the required relationship between the impact parameter b and the scattering angle θ.

The quantity X/mv_o^2 is related to the closest distance, D, which the particle could approach to the scatterer; this would occur during a head-on approach, scattering angle $\phi = 180^o$, $b = 0$; when equating the potential energy when the particle's velocity has dropped to zero to the initial kinetic energy, one obtains

$$\frac{X}{D} = \tfrac{1}{2}\, mv_o^2$$

showing

$$\frac{X}{mv_o^2} = \frac{D}{2}\,. \tag{5.55}$$

One can therefore rewrite eqns (5.53) and (5.54) as

$$b = \frac{D}{2} \cot\,(\phi/2). \tag{5.56}$$

The distance of closest approach can be of interest when b is not zero; here it is denoted as d (= OM) on Fig. 5.5. To calculate it, it is again not necessary to know the detailed geometry of the hyperbola, only to be able to apply the fundamental physics of the scattering process. The constancy of the scattered particles' total energy gives the equation

$$\tfrac{1}{2}\, mv^2 + \frac{X}{d} = \tfrac{1}{2}\, mv_o^2, \tag{5.57}$$

where v is the particles' speed at closest approach (point M), and the conservation of angular momentum gives

$$mvd = mv_o b. \tag{5.58}$$

Thus one has two simultaneous equations for v and d, and by eliminating v can obtain an expression for d. Suppose that it is desired to find d as a function of the scattering angle ϕ; this can be done as follows. Rearrangement of eqn (5.57) gives

$$\frac{v^2}{v_o^2} + \frac{2X}{dmv_o^2} = 1$$

which, from eqn (5.55), can be written as

$$\frac{v^2}{v_o^2} + \frac{D}{d} = 1. \qquad (5.59)$$

Substituting for (v^2/v_o^2) from eqn (5.58) and multiplying the result by d^2, one obtains the following quadratic equation for d:

$$d^2 - Dd - b^2 = 0.$$

Using the standard expression for the solution of a quadratic equation, it follows that

$$d = \tfrac{1}{2}\{D \pm (D^2 + 4b^2)^{\frac{1}{2}}\}$$

Substituting now for b from eqn (5.56) and rearranging, one gets

$$d = \frac{D}{2}\left[1 \pm \{1 + \cot^2(\phi/2)\}^{\frac{1}{2}}\right].$$

Remembering that $1 + \cot^2 \equiv \text{cosec}^2$ and that the cosecant of an angle is always equal to or greater than unity, it follows that, as d must be positive,

$$d = \frac{D}{2}\{1 + \text{cosec}\,(\phi/2)\}.$$

This provides the desired relationship between d and ϕ, showing that $d = D$ when $\phi = 180^o$, and $d \to \infty$ as $\phi \to 0$.

Rutherford's analysis of the experiments mentioned at the beginning of this section showed that the scattering of α-particles by gold atoms obeyed this theory with distances d and D much smaller than the estimated overall size of one atom. This led Rutherford to suggest that the atom must contain a positively-charged nucleus, much smaller than the atom itself, which would provide the necessary inverse square law scattering force. The final experiments were actually carried out by Geiger and Marsden in association with Rutherford; earlier non-nuclear atomic theory had suggested that there would be no measurable backward scattering as positive charge would be too diffusely spread through the atom.

5.7. CONCLUSION

In all the problems discussed in this chapter, the source of the central force has been treated as fixed in space; this has worked to a good approximation because the mass of the source has been large compared with the mass of the particle moving under the force. However, the source is itself free to move and is itself subjected to an opposite but equal force. The resulting motion of the two particles is such that the linear momentum of their centre of mass remains constant. There are a number of real cases when this becomes important, and one may wish to use centre of mass coordinates (Section 1.15) to deal with such problems. There tends to be more mathematics than basic physics in the treatment of central force problems, but this lies in the nature of the problem which can still be important physically; it underlines the need for the physicist to acquire a sound working background of mathematics, which is the language for much of his thinking.

5.8. EXAMPLES

1. A particle is moving in a circle under the action of a central force. What law relates the force to the radius of the orbit if (a) the orbit periods are found to be proportional to the square of the radius, (b) the orbit periods are found to be independent of the radius?

2. The planet Mars has a radius of 3400 km; its nearest moon,

Phobos, completes a circular orbit of radius 9300 km once every 7 h 39 min. From these data, calculate the acceleration due to gravity on the surface of Mars.

3. The distances from the sun of the earth and of Saturn are 149×10^6 km and 1426×10^6 km respectively. If the earth takes $365\frac{1}{4}$ days to complete one orbit, how long will Saturn take?

4. The minimum velocity with which an object would have to be projected into space from the surface of a planet sufficient to ensure that the object would never return under the action of the planet's gravitational force alone is known as the escape velocity for that planet. Given that the radius of the earth is 6400 km and that the acceleration due to gravity at the earth's surface is $9{\cdot}8$ $\mathrm{m\,s^{-1}}$, calculate the escape velocity for earth. If the law of gravitational force were an inverse nth power rather than an inverse square, but the earth's radius and g remained the same, would the escape velocity be any different?

5. A particle of mass m is moving under the influence of a central force of magnitude F given by

$$F = kr,$$

where r is the distance from the source of the force which is directed towards that source.
Show that the resultant motion is the sum of two SHMs of the same period in lines at right angles to one another, and discuss the possible movements of the particle.

6. A particle is moving under the influence of a force directed towards a point which obeys an inverse cube law with distance from the point. Examine the effect of this on eqns (5.41) and (5.42) and on the ease with which a

solution to them can be found.

7. A particle of mass m and velocity v_o enters a repulsive central force field F obeying the law

$$F = Xr^{-3},$$

where X is a constant. Calculate (a) the closest distance of approach D when the particle approaches the force centre head-on, and (b) the closest distance of approach d for the general case in terms of D and of the impact parameter b.

6. Analytical classical mechanics

6.1 INTRODUCTION

The basic classical mechanics, which is all that most physicists will require and use during their professional career, has already been surveyed in this book. This is a system of mechanics based on Newton's laws of motion, which are themselves deduced from experimental observations. It is adequate for the majority of problems which physicists are likely to encounter, and the principles which it contains are fundamental to all the basic physical theory of natural phenomena.

When Newton originally presented his ideas in 1687, his development of the theory was geometrical in its mathematical technique; as a result, he did not leave it in a generally useful form. Although his theory was quite clear as to the vector nature of forces, etc., the techniques for handling such quantities mathematically were not at that time effectively developed. The mechanics which he presented was also one describing the movement of particles rather than the movements and rotations of large rigid bodies; the extension of Newton's ideas to the latter class of problems was first achieved by L. Euler in 1736.

The concepts of energy and energy conservation were certainly not understood in Newton's time. However the quantity mv^2 (i.e., twice the kinetic energy, T) was studied and given the name *vis viva*. In 1751, P.L.M. de Maupertius put forward a principle known as the principle of least action. (Fermat's principle of least time had become a part of the theory of optics in Newton's time). This in essence stated that, between points s_1 and s_2 on a path or between times t_1 and t_2, a particle would move in such a way that

$$\int_{s_1}^{s_2} mv\,ds = \int_{t_1}^{t_2} 2T\,dt$$

would have a minimum value. This principle seemed to work, but it was left to J.L. Lagrange to show that it was consistent with Newton's laws of motion.

Lagrange was a mathematician of exceptional ability, whose great contribution to mechanics was to convert the geometrical mechanics of Newton into a usable algebraic form. He also introduced the concept of potential energy. His ideas on mechanics were published in 1788 in a book called *Mechanique Analytique*, and it is from the title of this book that the term analytical mechanics arises; it could equally well have been called algebraic mechanics. Lagrange, incidentally, was later largely responsible for the metric system of units (metre, gram, etc.) introduced in France in the early nineteenth century, and now used internationally for scientific work.

The potential energy concept allowed Lagrange to eliminate force from the laws of motion as the equation

$$\underline{F} = -\text{ grad } V$$

relating (in modern notation) force ($\underline{F}$) to potential energy (V) allows a substitution of potential energy gradient for force. His resulting equations of motion, Lagrange's equations, become most useful when extended to a system of generalized coordinates, and it is in problems requiring this extension for a solution that Lagrange's contribution to mechanics is perhaps most felt.

The ideas of Lagrange were further extended by W.R. Hamilton in 1834 and, when one now talks of analytical classical mechanics, one is really speaking about classical mechanics as developed by Lagrange and Hamilton.

6.2. LAGRANGE'S EQUATIONS OF MOTION

Consider a particle of mass m moving only in the x direction, so that its kinetic energy, $T = \frac{1}{2} m\dot{x}^2$, its momentum $m\dot{x}$ and any force F acting on it can be expressed in terms of the potential energy gradient by

$$F = -\frac{\partial V}{\partial x}. \qquad (6.1)$$

Now momentum and kinetic energy can be seen to be related by the equations

$$m\dot{x} = \frac{\partial}{\partial \dot{x}}(\tfrac{1}{2}m\dot{x}^2) = \frac{\partial T}{\partial \dot{x}}. \qquad (6.2)$$

Hence, from eqns (6.1) and (6.2), the equation of motion (rate of change of momentum equals force) for the particle can be written as

$$\frac{\mathrm{d}}{\mathrm{d}t}\left(\frac{\partial T}{\partial \dot{x}}\right) = -\frac{\partial V}{\partial x} \qquad (6.3)$$

or

$$\frac{\mathrm{d}}{\mathrm{d}t}\left(\frac{\partial T}{\partial \dot{x}}\right) + \frac{\partial V}{\partial x} = 0. \qquad (6.4)$$

In his treatment of mechanics, Lagrange introduced a function L, now called the Lagrangian, defined by

$$L = T - V. \qquad (6.5)$$

As, in this case, $T = T(\dot{x})$, and if also $V = V(x)$, it immediately follows that

$$\frac{\partial L}{\partial \dot{x}} = \frac{\partial T}{\partial x} \text{ and } \frac{\partial L}{\partial x} = -\frac{\partial V}{\partial x}, \qquad (6.6)$$

so that eqn (6.4) becomes

$$\frac{\mathrm{d}}{\mathrm{d}t}\left(\frac{\partial L}{\partial \dot{x}}\right) - \frac{\partial L}{\partial x} = 0. \qquad (6.7)$$

Eqn (6.7) is the equation of motion written in the Lagrangian form. Similar equations can be written for the y and z coordinates and the three equations so produced will apply to motion in a three-dimensional space described by the coordinates x, y, z. In writing them down in this way, no new concepts have been introduced and the introduction of the Lagrangian function has added little of obvious usefulness. For very simple problems, Lagrange's equations of motion can merely provide an elaborate way of reaching the desired solution. As already stated, where Lagrange made a significant advance was in the development of the concept of generalized coordinates, following up ideas introduced by Euler. This topic will be discussed briefly in the next section.

6.3. CONCEPT OF GENERALIZED COORDINATES

The important thing about Lagrange's equations of motion is that they hold not only in cartesian coordinates x, y, z, but in any useful set of coordinates. An example of useful coordinates is provided by polar coordinates, r, θ, ϕ. The coordinates likely to be useful depend on the problem to be tackled; for example, if one wishes to establish the motion of two masses hung from a light cord which passes over a frictionless support, as is shown in Fig. 6.1, the height of one of the masses from the ground, x, will suffice as an adequate coordinate. The number of coordinates required will equal the number of degrees of freedom available for the motion to be described. Let there be n coordinates, q_1, q_2 ------ q_n, and let i be an integer such that $1 \leqslant i \leqslant n$. Then Lagrange's equations of motion will take the form of i different equations, each of the form

$$\frac{\mathrm{d}}{\mathrm{d}t}\left(\frac{\partial L}{\partial \dot{q}_i}\right) - \frac{\partial L}{\partial q_i} = 0. \qquad (6.8)$$

A rigorous proof of this statement is beyond the scope of this

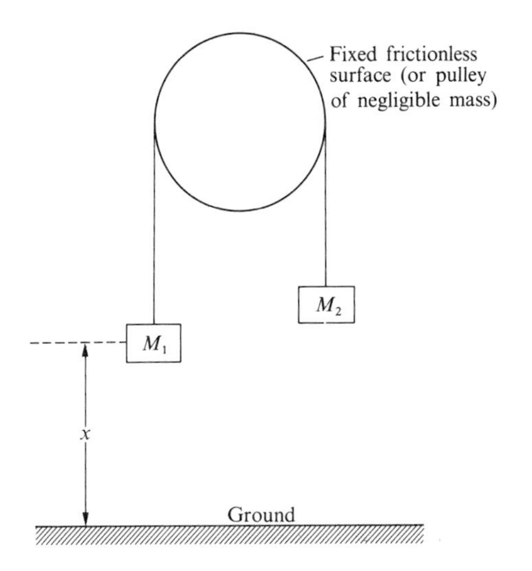

FIG. 6.1. To illustrate a generalized coordinate; the coordinate x is sufficient to locate M_1 and M_2.

book; the deduction in Section 6.2 is restricted to cartesian coordinates.

As an example, consider how eqn (6.8) can be applied to the problem of Fig. 6.1, using the generalized coordinate x. First, the Lagrangian, L, of the system consisting of the two masses M_1 and M_2, is given by

$$L = T - V \tag{6.9}$$

where the kinetic energy

$$T = \tfrac{1}{2}\,(M_1 + M_2)\,\dot{x}^2, \tag{6.10}$$

and the potential energy

$$V = M_1 g x + M_2 g(A - x) \tag{6.11}$$

A being a constant equal to the height of M_2 when M_1 is on the ground ($x = 0$). Thus, from eqns (6.9), (6.10), and (6.11),

$$L = \tfrac{1}{2}\,(M_1 + M_2)\dot{x}^2 - \{M_1 gx + M_2 g(A - x)\}. \qquad (6.12)$$

One can now calculate the terms of eqn (6.8). First

$$\frac{\partial L}{\partial \dot{x}} = (M_1 + M_2)\dot{x},$$

$$\frac{\mathrm{d}}{\mathrm{d}t}\left(\frac{\partial L}{\partial \dot{x}}\right) = (M_1 + M_2)\ddot{x}; \qquad (6.13)$$

also

$$\frac{\partial L}{\partial x} = -\,(M_1 - M_2)g. \qquad (6.14)$$

Thus, from eqns (6.13) and (6.14), Lagrange's equation of motion, eqn (6.8), takes the form

$$(M_1 + M_2)\ddot{x} + (M_1 - M_2)g = 0, \qquad (6.15)$$

giving the result

$$\ddot{x} = \frac{M_2 - M_1}{M_1 + M_2}\,g \qquad (6.16)$$

for the acceleration of the system described in terms of the variation of the generalized coordinate x.

Dealing with a simple problem of this type by means of Lagrange's equations is rather like taking a sledgehammer to crack a nut; but it is however instructive. It shows that the problem is soluble without mentioning the word force, provided one has an expression for the potential. It also shows that the problem is soluble without reference to the internal force, i.e., the tension in the string. It also shows that if one wished to know the tension in the string, one would not use this method of approach.

Eqn (6.13) is also instructive. Referring back first to Section 6.2, eqns (6.2) and (6.6) make it clear that, in the problem discussed there, $(\partial L/\partial \dot{x})$ was the real momentum of the particle. However, the quantity given by eqn (6.13) is not a real momentum; $(M_1 + M_2)\dot{x}$ is not the real momentum of M_1 and M_2 together, for M_1 has momentum $M_1\dot{x}$ upwards while M_2 has momentum $M_2\dot{x}$ downwards. The real upward momentum of M_1 and M_2 together is $(M_1 - M_2)\dot{x}$. However $(M_1 + M_2)\dot{x}$ has been seen to behave very like a real momentum and is an example of what is known as a generalized momentum. In fact, the quantity $(\partial L/\partial \dot{q}_i)$ is often known as the generalized momentum.

The central force problem provides another useful example of the application of Lagrange's equations. Consider a particle of mass m moving in a plane, coordinates r and θ, under the action of a central force from the origin such that the potential energy, V, of the particle is $-(k/r)$. [Eqn (5.3) was a special case of this with $k = GMm$]

Using eqns (5.29) and (6.5) the Lagrangian takes the form

$$L = \tfrac{1}{2}\, m(\dot{r}^2 + r^2\,\dot{\theta}^2) + \frac{k}{r}\,. \tag{6.17}$$

The generalized coordinates are r and θ, so Lagrange's equations of motion for this problem are

$$\frac{\mathrm{d}}{\mathrm{d}t}\left(\frac{\partial L}{\partial \dot{r}}\right) - \frac{\partial L}{\partial r} = 0 \tag{6.18}$$

and

$$\frac{\mathrm{d}}{\mathrm{d}t}\left(\frac{\partial L}{\partial \dot{\theta}}\right) - \frac{\partial L}{\partial \theta} = 0. \tag{6.19}$$

From eqn (6.17), eqn (6.18) becomes

$$m\ddot{r} - mr\dot{\theta} + \frac{k}{r^2} = 0. \tag{6.20}$$

Similarly, eqn (6.19) becomes

$$\frac{\mathrm{d}}{\mathrm{d}t}\,(mr^2\dot{\theta}) = 0. \tag{6.21}$$

Consider in turn the last two equations. From the fact that $mr^2\dot{\theta}$ is the angular momentum of m about the origin, eqn (6.21) is at once recognizable as a statement that the angular momentum is constant, which is a known property of motion under a central force; integration with respect to time produces eqn (5.36) [or eqn (5.4)]. Eqn (6.20) is recognizable too as a form of eqn (5.37). Thus the eqns (6.20) and (6.21) provide a suitable starting point for this central orbit problem.

Eqn (6.20) can be rearranged in the form

$$mr\dot{\theta}^2 - \frac{k}{r^2} = \frac{\mathrm{d}}{\mathrm{d}t}(m\dot{r}), \tag{6.22}$$

which shows a marked analogy to the Newtonian equation of motion 'Force = rate of change of momentum'. On the right, the equation has the rate of change of generalized momentum $m\dot{r}$; on the left, the first term is the centrifugal force, an inertial reaction rather than a true force (see Section 2.3), while the second term is a real force derivable from a potential. This again illustrates how the use of generalized coordinates produces a system of mechanics mathematically comparable with the Newtonian system, although the physics behind it is not always as immediately clear.

6.4. THE GENERALIZED FORCE

The discussion of Lagrangian mechanics in Sections 6.2 and 6.3. has been confined to movement under the action of a force derivable from a potential energy or potential gradient; but a moving mass m can experience other forces which are not derivable from potentials. If, for example, one pushes a piece of furniture sideways, the force one is exerting to accelerate the furniture is not in any simple way derivable from a potential which is single-valued in space. Now consider eqn (6.3), i.e.,

$$\frac{\mathrm{d}}{\mathrm{d}t}\left(\frac{\partial T}{\partial \dot{x}}\right) = -\frac{\partial V}{\partial x} \tag{6.23}$$

Suppose there was some other force, Q, acting in the $+x$ direction. Eqn (6.23) would then become, for the problem under discussion in Section 6.2,

$$\frac{\mathrm{d}}{\mathrm{d}t}\left(\frac{\partial T}{\partial \dot{x}}\right) = -\frac{\partial V}{\partial x} + Q.$$

The argument would continue as before and, in place of eqn (6.7), one would obtain

$$\frac{\mathrm{d}}{\mathrm{d}t}\left(\frac{\partial L}{\partial \dot{x}}\right) - \frac{\partial L}{\partial x} = Q. \qquad (6.24)$$

In this case, Q is a true force. However, this treatment of an extra force can be extended to the case of generalized coordinates and in place of eqn (6.8), one can write

$$\frac{\mathrm{d}}{\mathrm{d}t}\left(\frac{\partial L}{\partial \dot{q}_i}\right) - \frac{\partial L}{\partial q_i} = Q_i$$

where Q_i is now the component of the generalized force, Q, appropriate to the generalized coordinate, q_i. This generalized force component will be such that the product $Q_i \mathrm{d}q_i$ represents work.

To illustrate this last point, let it be supposed that an extra force Q has to be added to the central force problem so that eqn (6.19) has to be rewritten as

$$\frac{\mathrm{d}}{\mathrm{d}t}\left(\frac{\partial L}{\partial \dot{\theta}}\right) - \frac{\partial L}{\partial \theta} = Q_\theta,$$

where Q_θ is the appropriate component of this generalized force. In place of eqn (6.21), one now obtains

$$\frac{\mathrm{d}}{\mathrm{d}t}(mr^2\dot{\theta}) = Q_\theta,$$

i.e., Q_θ equals the rate of change of angular momentum. Thus [see eqn (2.23)] the generalized force Q has a component Q_θ, which is a torque rather than a true force. The rate of change of the generalized momentum has equalled the appropriate generalized force component, and if Q_θ acts through $d\theta$, the work done is $Q_\theta \, d\theta$.

6.5. HAMILTON'S PRINCIPLE

Sir William Rowan Hamilton had in 1827 the remarkable distinction of being appointed Professor of Astronomy at Trinity College, Dublin, while he was still an undergraduate there. His studies in analytical mechanics were built on the foundations established by Lagrange. Often quoted, and occasionally confused with the principle of least action, is Hamilton's principle. This can be stated as follows: between times t_1 and t_2, a particle will move in such a way that

$$\delta \int_{t_1}^{t_2} L \, dt = 0.$$

This means that the integral has a stationary value, usually in practice a minimum, as the path is varied. It is of course clearly closely related to the principle of least action and, like that principle, can be derived from the fundamental mechanical laws, Newton's laws of motion. It can be used as a starting point for the formal development of analytical mechanics, but no attempt will be made to do so here.

6.6. THE HAMILTONIAN

In the Lagrangian formulation of mechanics, the variables describing the motion of a particle were $\dot{q}_i$, q_i and t; see for example eqn (6.8). Hamilton chose instead to use the variables p_i, q_i and t, where p_i is the generalized momentum associated with the coordinate q_i and defined by

$$p_i = \frac{\partial L}{\partial \dot{q}_i} . \tag{6.25}$$

He then introduced a function H, now known as the Hamiltonian, defined by

$$H = \Sigma\ p_i \dot{q}_i - L, \tag{6.26}$$

where the summation is taken over all the necessary values of i.

Consider now the Hamiltonian for the simple case discussed in Section 6.2, where there is only one coordinate to consider. For this case

$$\begin{aligned} \Sigma\ p_i \dot{q}_i &= m\dot{x}\cdot\dot{x} \\ &= 2T, \end{aligned}$$

and hence

$$\begin{aligned} H &= 2T - L \\ &= 2T - (T - V), \end{aligned}$$

i.e.,

$$H = T + V. \tag{6.27}$$

If one takes any case of motion in a conservative force field, where the potential energy is a function of position only, eqn (6.27) will be found to hold true. Thus, if the total energy is E, one can write

$$H = E. \tag{6.28}$$

For many practical problems, eqn (6.28) can be usefully applied to establish the form of the Hamiltonian. However, there are circumstances in which the Hamiltonian is not the total energy, so care must be exercised in using this equality. It is not always necessary that either H or E or both should remain constant throughout a motion, and cases occur where $H = E$ but both vary. Should external forces appear, or should the potential become a function of time or velocity, then eqn (6.28) may no longer be true.

6.7. HAMILTON'S CANONICAL EQUATIONS

As

$$H = H\ (p_i, q_i, t),$$

it follows that

$$dH = \sum_i \frac{\partial H}{\partial p_i}\, dp_i + \sum_i \frac{\partial H}{\partial q_i}\, dq_i + \frac{\partial H}{\partial t}\, dt. \qquad (6.29)$$

However, the Hamiltonian is defined by eqn (6.26) to be

$$H = \sum_i p_i\, d\dot{q}_i - L\ (\dot{q}_i, q_i, t),$$

so that

$$dH = \sum_i p_i\, d\dot{q}_i + \sum_i \dot{q}_i dp_i - \sum_i \frac{\partial L}{\partial \dot{q}_i}\, d\dot{q}_i - \sum_i \frac{\partial L}{\partial q_i}\, dq_i - \frac{\partial L}{\partial t}\, dt. \qquad (6.30)$$

From eqn (6.25) one can see at once that the first and third terms on the r.h.s. of eqn (6.30) cancel. Comparison of the remaining terms with the corresponding terms of eqn (6.29) shows that

$$\frac{\partial H}{\partial p_i} = \dot{q}_i, \qquad (6.31)$$

and also

$$\frac{\partial H}{\partial q_i} = - \frac{\partial L}{\partial q_i}. \qquad (6.32)$$

Eqns (6.8) and (6.25) allow one to substitute $\dot{p}_i$ for $(\ L/\ q_i)$, so

$$\frac{\partial H}{\partial q_i} = - \dot{p}_i \qquad (6.33)$$

Eqns (6.31) and (6.33) are known as Hamilton's canonical equations of motion. While it is possible to use them in the solution of elementary problems, and for instructional purposes it may be desirable to do so, their real value does not lie there but in the more advanced theory of mechanics which arises from them.

6.8. CONCLUSION

This brief discussion of analytical mechanics has introduced certain methods of formal approach which find application in the treatment of otherwise difficult problems in physics. No new fundamental concepts have been involved; as Mach pointed out, the aim of analytical mechanics is the mastery of problems. Certain terms find their way beyond classical mechanics; the Hamiltonian, for example, is encountered in quantum mechanics, and here one has only dealt briefly with its classical background. The fundamental concepts of mechanics are what really matter, but they will not be of use unless they can be applied; for many problems, analytical mechanics provides the tools for such application.

6.9. EXAMPLES

1. A particle of mass m is moving in a straight line in such a way that, when it is a distance x from a point in that line, its potential energy is $\frac{1}{2}kx^2$ where k is a constant. Write down the Lagrangian for the motion and hence show that the equation of motion for this case written in the Lagrangian form is identical to the standard equation for SHM.
2. A frictionless pulley system has one pulley suspended by its axis from a rigid support. A cord round this pulley is attached to a mass of magnitude $4m$ at one end and to the axis of a second pulley at the other. A cord round this second pulley is attached to masses of magnitude $2m$ and m, one at either end. The whole system is free to move.

Show that, if the masses of the pulleys and cords can be neglected, the downward acceleration of the mass of magnitude $4m$ is $g/5$, and calculate the accelerations of the other two masses. In solving this problem, use Lagrange's equations of motion; two suitable generalized coordinates are the distance of the first mass from the ground and the distance of either of the other two masses from the pulley suspending it; however, these are not the only suitable coordinates.

3. A uniform wheel of radius r is free to rotate about a horizontal axis through its centre and perpendicular to its plane; its moment of inertia about this axis is I. A mass m is fixed to its edge and it is set spinning anticlockwise with angular velocity $\dot{\theta}$. Assume that $\theta = 0$ when the mass is moving vertically upwards. Write down (a) the Lagrangian for the motion, and (b) Lagrange's equation of motion for the generalized coordinate θ, showing that it takes the form

$$(I + mr^2)\ddot{\theta} + mgr \cos \theta = 0.$$

Examine how this result could have been achieved without using the Lagrangian method. Compare any assumptions which may have been made in the two approaches to the problem.

4. Consider the motion of question 6.1. Use eqn (6.26) to write down the Hamiltonian. Confirm that eqn (6.27) follows in this case. Write down also Hamilton's canonical equations for this motion and show that one of them is the standard equation for SHM.

5. Consider a particle of mass m moving in a plane under the action of a central force F directed towards the origin of polar coordinates r, θ and of magnitude

$$F = kr^{-2}$$

Show that the Hamiltonian for the motion can be expressed as

$$H = \frac{p_r^2}{2m} + \frac{p_\theta^2}{2mr^2} - \frac{k}{r} ,$$

where p_r and p_θ are the generalized momenta appropriate to the coordinates r and θ respectively. Examine and comment upon the four Hamilton's canonical equations of motion for this case.

Answers to examples

CHAPTER 1

3. 3 min 24 s; 1067 kW. 4. Ft/M. 5. $X\rho v/t$; ρv^2.
8. 4.8×10^{-4}. 9. (a) $Md(g/l)^{\frac{1}{2}}/2m$; (b) $(M - m)/M$. 10. m.

CHAPTER 2.

5. 1/3. 9. $Ma^2/6$; $Ma^2/6$; $Ma^2/6$.
10. (a) (i) 16 N (1.8 kg), (ii) 240 N (24.5 kg); (b) (i) and (ii) 222 N (22.7 kg).

CHAPTER 3.

1. (a) 4.3 s; (b) 4.1 min. 9. At velocity resonance.

CHAPTER 4.

3. 5.38 N. 4. 2.97×10^3 m s^{-1}. 5. 0.139 nepers m^{-1}; 1.20 dB m^{-1}. 7. (a) c^2/v; (b) v.

CHAPTER 5.

1. (a) $F \propto r^{-3}$; (b) $F \propto r$. 2. 3.62 m s^{-2}. 3. 1.081×10^{-4} days. 4. 11.2 km s^{-1}; no. 7. $(X/mv_o^2)^{\frac{1}{2}}$; $(D^2 + b^2)^{\frac{1}{2}}$.

CHAPTER 6.

2. $2m$: $g/5$ down; m: $3g/5$ up.

Index

Physical constants and conversion factors

Avogadro constant	L or N_A	$6{\cdot}022 \times 10^{23}$ mol^{-1}
Bohr magneton	μ_B	$9{\cdot}274 \times 10^{-24}$ J T^{-1}
Bohr radius	a_0	$5{\cdot}292 \times 10^{-11}$ m
Boltzmann constant	k	$1{\cdot}381 \times 10^{-23}$ J K^{-1}
charge of an electron	e	$-1{\cdot}602 \times 10^{-19}$ C
Compton wavelength of electron	$\lambda_C = h/m_e c =$	$2{\cdot}426 \times 10^{-12}$ m
Faraday constant	F	$9{\cdot}649 \times 10^4$ C mol^{-1}
fine structure constant	$\alpha = \mu_0 e^2 c/2h =$	$7{\cdot}297 \times 10^{-3}$ ($\alpha^{-1} = 137{\cdot}0$)
gas constant	R	$8{\cdot}314$ J K^{-1} mol^{-1}
gravitational constant	G	$6{\cdot}673 \times 10^{-11}$ N m^2 kg^{-2}
nuclear magneton	μ_N	$5{\cdot}051 \times 10^{-27}$ J T^{-1}
permeability of a vacuum	μ_0	$4\pi \times 10^{-7}$ H m^{-1} exactly
permittivity of a vacuum	ϵ_0	$8{\cdot}854 \times 10^{-12}$ F m^{-1} ($1/4\pi\epsilon_0 = 8{\cdot}988 \times 10^9$ m F^{-1})
Planck constant	h	$6{\cdot}626 \times 10^{-34}$ J s
(Planck constant)/2π	$\hbar$	$1{\cdot}055 \times 10^{-34}$ J s $= 6{\cdot}582 \times 10^{-16}$ eV s
rest mass of electron	m_e	$9{\cdot}110 \times 10^{-31}$ kg $= 0{\cdot}511$ MeV/c^2
rest mass of proton	m_p	$1{\cdot}673 \times 10^{-27}$ kg $= 938{\cdot}3$ MeV/c^2
Rydberg constant	$R_\infty = \mu_0^2 m_e e^4 c^3/8h^3 =$	$1{\cdot}097 \times 10^7$ m^{-1}
speed of light in a vacuum	c	$2{\cdot}998 \times 10^8$ m s^{-1}
Stefan–Boltzmann constant	$\sigma = 2\pi^5 k^4/15h^3c^2 =$	$5{\cdot}670 \times 10^{-8}$ W m^{-2} K^{-4}
unified atomic mass unit (^{12}C)	u	$1{\cdot}661 \times 10^{-27}$ kg $= 931{\cdot}5$ MeV/c^2
wavelength of a 1 eV photon		$1{\cdot}243 \times 10^{-6}$ m

1 Å $= 10^{-10}$ m; 1 dyne $= 10^{-5}$ N; 1 gauss (G) $= 10^{-4}$ tesla (T);
0°C $= 273{\cdot}15$ K; 1 curie (Ci) $= 3{\cdot}7 \times 10^{10}$ s^{-1};
1 J $= 10^7$ erg $= 6{\cdot}241 \times 10^{18}$ eV; 1 eV $= 1{\cdot}602 \times 10^{-19}$ J; 1 cal$_{th}$ $= 4{\cdot}184$ J;
$\ln 10 = 2{\cdot}303$; $\ln x = 2{\cdot}303 \log x$; $e = 2{\cdot}718$; $\log e = 0{\cdot}4343$; $\pi = 3{\cdot}142$